Lecture Notes in Mathematics

Edited by J.-M. Morel, F. Takens and B. Teissier

Editorial Policy
for the publication of monographs

1. Lecture Notes aim to report new developments in all areas of mathematics and their applications – quickly, informally and at a high level. Mathematical texts analysing new developments in modelling and numerical simulation are welcome.

   Monograph manuscripts should be reasonably self-contained and rounded off. Thus they may, and often will, present not only results of the author but also related work by other people. They may be based on specialised lecture courses. Furthermore, the manuscripts should provide sufficient motivation, examples and applications. This clearly distinguishes Lecture Notes from journal articles or technical reports which normally are very concise. Articles intended for a journal but too long to be accepted by most journals, usually do not have this "lecture notes" character. For similar reasons it is unusual for doctoral theses to be accepted for the Lecture Notes series, though habilitation theses may be appropriate.

2. Manuscripts should be submitted (preferably in duplicate) either to one of the series editors or to Springer-Verlag, Heidelberg. In general, manuscripts will be sent out to 2 external referees for evaluation. If a decision cannot yet be reached on the basis of the first 2 reports, further referees may be contacted: The author will be informed of this. A final decision to publish can be made only on the basis of the complete manuscript, however a refereeing process leading to a preliminary decision can be based on a pre-final or incomplete manuscript. The strict minimum amount of material that will be considered should include a detailed outline describing the planned contents of each chapter, a bibliography and several sample chapters.

   Authors should be aware that incomplete or insufficiently close to final manuscripts almost always result in longer refereeing times and nevertheless unclear referees' recommendations, making further refereeing of a final draft necessary. Authors should also be aware that parallel submission of their manuscript to another publisher while under consideration for LNM will in general lead to immediate rejection.

3. Manuscripts should in general be submitted in English. Final manuscripts should contain at least 100 pages of mathematical text and should include
   - a table of contents;
   - an informative introduction, with adequate motivation and perhaps some historical remarks: it should be accessible to a reader not intimately familiar with the topic treated;
   - a subject index: as a rule this is genuinely helpful for the reader.

Continued on inside back-cover

Lecture Notes in Mathematics 1818

Editors:
J.-M. Morel, Cachan
F. Takens, Groningen
B. Teissier, Paris

Springer
Berlin
Heidelberg
New York
Hong Kong
London
Milan
Paris
Tokyo

Michael Bildhauer

Convex Variational Problems

Linear, Nearly Linear and Anisotropic Growth Conditions

 Springer

Author

Michael Bildhauer
Department of Mathematics
Saarland University
P.O. Box 151150
66041 Saarbrücken
Germany
e-mail: bibi@math.uni-sb.de

Cataloging-in-Publication Data applied for
Bibliographic information published by Die Deutsche Bibliothek

Die Deutsche Bibliothek lists this publication in the Deutsche Nationalbibliografie;
detailed bibliographic data is available in the Internet at http://dnb.ddb.de

Mathematics Subject Classification (2000): 49-02, 49N60, 49N15, 35-02, 35J20, 35J50

ISSN 0075-8434
ISBN 3-540-40298-5 Springer-Verlag Berlin Heidelberg New York

Springer-Verlag Berlin Heidelberg New York a member of BertelsmannSpringer
Science + Business Media GmbH

http://www.springer.de

© Springer-Verlag Berlin Heidelberg 2003
Printed in Germany

Typesetting: Camera-ready TeX output by the authors

SPIN: 10935590   41/3142/du - 543210 - Printed on acid-free paper

Dedicated to *Christina*

Preface

In recent years, two (at first glance) quite different fields of mathematical interest have attracted my attention.

- Elliptic variational problems with linear growth conditions. Here the notion of a "solution" is not obvious and, in fact, the point of view has to be changed several times in order to get some deeper insight.
- The study of the smoothness properties of solutions to convex anisotropic variational problems with superlinear growth.

It took some time to realize that, in spite of the fundamental differences and with the help of some suitable theorems on the existence and uniqueness of solutions in the case of linear growth conditions, a non-uniform ellipticity condition serves as the main tool towards a unified view of the regularity theory for both kinds of problems.

This is roughly speaking the background of my habilitations thesis at the Saarland University which is the basis for this presentation.

Of course there is a long list of people who have contributed to this monograph in one or the other way and I express my thanks to each of them. Without trying to list them all, I really want to mention:

Prof. G. Mingione is one of the authors of the joint paper [BFM]. The valuable discussions on variational problems with non-standard growth conditions go much beyond this publication.

Prof. G. Seregin took this part in the case of variational problems with linear growth.

Large parts of the presented material are joint work with Prof. M. Fuchs: this, in the best possible sense, requires no further comment. Moreover, I am deeply grateful for the numerous discussions and the helpful suggestions.

Saarbrücken, April 2003   *Michael Bildhauer*

Preface

Contents

1

Introduction

One of the most fundamental problems arising in the calculus of variations is to minimize strictly convex energy functionals with respect to prescribed Dirichlet boundary data. Numerous applications for this type of variational problems are found, for instance, in mathematical physics or geometry.

Here we do not want to give an introduction to this topic – we just refer to the monograph of Giaquinta and Hildebrandt ([GH]), where the reader will find in addition an intensive discussion of historical facts, examples and references.

Let us start with a more precise formulation of the problem under consideration: given a bounded Lipschitz domain $\Omega \subset \mathbb{R}^n$, $n \geq 2$, and a variational integrand $f\colon \mathbb{R}^{nN} \to \mathbb{R}$ of class $C^2(\mathbb{R}^{nN})$ we consider the autonomous minimization problem

$$J[w] := \int_\Omega f(\nabla w) \, \mathrm{d}x \longrightarrow \min \qquad (\mathcal{P})$$

among mappings $w\colon \Omega \to \mathbb{R}^N$, $N \geq 1$, with prescribed Dirichlet boundary data u_0. Depending on f, the comparison functions are additionally assumed to be elements of a suitable energy class $\mathbb{K}$. In the following, the variational integrand is always assumed to be strictly convex (in the sense of definition), thus we do not touch the quasiconvex case (compare, for instance, [Ev], [FH], [EG1], [AF1], [AF2], [CFM]).

The purpose of our studies is to establish regularity results for (maybe generalized and not necessarily unique) minimizers of the problem $(\mathcal{P})$ under linear, nearly linear and/or anisotropic growth conditions on f together with some appropriate notion of ellipticity: if u denotes a suitable (weak) solution of $(\mathcal{P})$, then three different kinds of results are expected to be true.

THEOREM 1 (Regularity in the scalar case)
Assume that $N = 1$ and that f satisfies some appropriate growth and ellipticity conditions. Then u is of class $C^{1,\alpha}(\Omega)$ for any $0 < \alpha < 1$.

According to an example of DeGiorgi (see [DG3], compare also [GiuM2], [Ne] and the recent example [SY]), there is no hope to prove an analogous result of this strength in the vectorial setting. Here we can only hope for

THEOREM 2 (PARTIAL REGULARITY IN THE VECTOR-VALUED CASE)
Assume that $N > 1$ and that f satisfies some appropriate growth and ellipticity conditions. Then there is an open set $\Omega_0 \subset \Omega$ of full Lebesgue measure, i.e. $|\Omega - \Omega_0| = 0$, such that $u \in C^{1,\alpha}(\Omega_0; \mathbb{R}^N)$, $0 < \alpha < 1$.

Finally, an additional structure condition might improve Theorem 2 to full regularity (see [Uh], earlier ideas are due to [Ur]):

THEOREM 3 (FULL REGULARITY IN THE VECTOR-VALUED CASE WITH SOME ADDITIONAL STRUCTURE)
Suppose that in the vectorial setting the integrand f satisfies in addition $f(Z) = g(|Z|^2)$ for some function $g: [0,\infty) \to [0,\infty)$ of class C^2 (plus some Hölder condition for the second derivatives). Then u is of class $C^{1,\alpha}(\Omega; \mathbb{R}^N)$, $0 < \alpha < 1$.

As the essential assumptions, the growth and the ellipticity conditions on f are involved in the above theorems. Hence, in order to make our discussion more precise and to summarize the various cases for which Theorems 1–3 are known to be true, we first introduce some brief classification of the integrands under consideration with respect to both growth and ellipticity properties. We also remark that in the cases A and B considered below the existence (and the uniqueness) of minimizers in suitable energy spaces is easily established.

Before going through the following list it should be emphasized that we do not claim to give an historical overview which is complete to some extent.

A.1 POWER GROWTH

Having the standard example $f_p(Z) = (1 + |Z|^2)^{p/2}$, $1 < p$, in mind, let us assume that the growth rates from above and below coincide, i.e. for some number $p > 1$ and with constants c_1, c_2, C, λ, $\Lambda > 0$ the integrand f satisfies for all $Z, Y \in \mathbb{R}^{nN}$ (note that the second line of (1) implies the first one)

$$c_1 |Z|^p - c_2 \leq \quad f(Z) \quad \leq C\left(1 + |Z|^p\right),$$
$$\lambda\left(1 + |Z|^2\right)^{\frac{p-2}{2}} |Y|^2 \leq D^2 f(Z)(Y,Y) \leq \Lambda\left(1 + |Z|^2\right)^{\frac{p-2}{2}} |Y|^2. \tag{1}$$

With the pioneering work of DeGiorgi, Moser, Nash as well as of Ladyzhenskaya and Ural'tseva, Theorem 1 is well known in this setting and of course many other authors could be mentioned (see [DG1], [Mos], [Na] and [LU1] for a complete overview and a detailed list of references).

As already noted above, the third theorem in this setting should be mainly connected to the name of Uhlenbeck (see [Uh], where the full strength of (1) is not needed which means that also degenerate ellipticity can be considered).

Without additional structure conditions in the vectorial case, the two-dimensional case $n = 2$ substantially differs from the situation in higher dimensions: a classical result of Morrey ensures full regularity if $n = 2$ (here we like to refer to [Mor1], the first monograph on multiple integrals in the calculus of variations, where again detailed references can be found).

Finally, Theorem 2 is proved in any dimension and in a quite general setting by Anzellotti/Giaquinta ([AG2]), where the whole scale of integrands up to the limit case of linear growth is covered (with some suitable notion of relaxation). In addition, the assumptions on the second derivatives are much weaker than stated above, i.e. their partial regularity result is true whenever $D^2 f(Z) > 0$ holds for any matrix Z.

To keep the historical line, we like to mention the earlier contributions on partial regularity [Mor2], [GiuM1], [Giu1] (compare also [DG2], [Alm], a detailed overview is found in [Gia1]).

A.2 ANISOTROPIC POWER GROWTH

The study of anisotropic variational problems was pushed by Marcellini ([Ma2]–[Ma7]) and is a natural extension of (1). To give some motivation we consider the case $n = 2$, $2 \leq p \leq q$ and replace f_p by

$$f_{p,q}(Z) = \left(1 + |Z|^2\right)^{\frac{p}{2}} + \left(1 + |Z_2|^2\right)^{\frac{q}{2}} , \quad Z = (Z_1, Z_2) \in \mathbb{R}^{2N} ,$$

hence f is allowed to have different growth rates from above and from below. The natural generalization of the structure condition (1) is the requirement that f satisfies (again the growth conditions on the second derivatives imply the corresponding growth rates of f)

$$c_1 |Z|^p - c_2 \leq \quad f(Z) \quad \leq C\left(1 + |Z|^q\right) ,$$
$$\lambda \left(1 + |Z|^2\right)^{\frac{p-2}{2}} |Y|^2 \leq D^2 f(Z)(Y, Y) \leq \Lambda \left(1 + |Z|^2\right)^{\frac{q-2}{2}} |Y|^2$$

(2)

for all Z, $Y \in \mathbb{R}^{nN}$, where, as usual, c_1, c_2, C, λ, Λ denote some positive constants and $1 < p \leq q$.

If p and q differ too much, then it turns out that even in the scalar case singularities may occur (to mention only one famous example we refer to [Gia2]). However, following the work of Marcellini, suitable assumptions on p and q yield regular solutions (compare Section 3.5 for a discussion of these conditions). Note that [Ma5] also covers the case $N > 1$ with some additional structure condition.

In the general vectorial setting only a few contributions are available, we like to refer to the papers of Acerbi/Fusco ([AF4]) and Passarelli Di Napoli/Siepe ([PS]), where partial regularity results are obtained under quite restrictive assumptions on p and q excluding any subquadratic growth (again see Section 3.5).

If an additional boundedness condition is imposed, then the above results are improved by Esposito/Leonetti/Mingione ([ELM2]) and Choe ([Ch]). In [ELM2] higher integrability (up to a certain extent) is established ($N \geq 1$, $2 \leq p$) under a quite weak relation between p and q. A theorem of the third type is found in [Ch].

B.1 GROWTH CONDITIONS INVOLVING N-FUNCTIONS

Studying the monograph of Fuchs and Seregin ([FuS2]) it is obvious that many problems in mathematical physics are not within the reach of power growth models – the theories of Prandtl-Eyring fluids and of plastic materials with logarithmic hardening serve as typical examples. The variational integrands under consideration are now of nearly linear growth, for example we have to study the logarithmic integrand

$$f(Z) = |Z| \ln(1 + |Z|)$$

which satisfies none of the conditions (1) or (2).

The main results on integrands with logarithmic structure are proved by Frehse/Seregin ([FrS]: full regularity if $n = 2$), Fuchs/Seregin ([FuS1]: partial regularity if $n \leq 4$), Esposito/Mingione ([EM2]: partial regularity in any dimension) and finally by Mingione/Siepe ([MS]: full regularity in any dimension).

B.2 THE FIRST EXTENSION OF THE LOGARITHM

As a first natural extension one may think of integrands which are bounded from above and below by the same quantity $A(|Z|)$, where $A: [0, \infty) \to [0, \infty)$ denotes some arbitrary N-function satisfying a Δ_2-condition (see [Ad] for the precise definitions). Although this does not imply some natural bounds (in terms of A) on the second derivatives, (1) and (2) suggest the following model: given a N-function A as above, positive constants c, C, λ and Λ, we assume that our integrand f satisfies

$$cA(|Z|) \leq \quad f(Z) \quad \leq C A(|Z|) \,,$$
$$\lambda \left(1 + |Z|^2\right)^{-\frac{\mu}{2}} |Y|^2 \leq D^2 f(Z)(Y, Y) \leq \Lambda \left(1 + |Z|^2\right)^{\frac{q-2}{2}} |Y|^2 \tag{3}$$

for all Z, $Y \in \mathbb{R}^{nN}$ and for some real numbers $1 \leq \mu$, $1 < q \leq 2$, this choice being adapted to the logarithmic integrand which satisfies (3) with $\mu = 1$ and $q = 1 + \varepsilon$ for any $\varepsilon > 0$. Note that the correspondence to (1) and (2) is only of formal nature: since we require $\mu \geq 1$, the μ-ellipticity condition, i.e. the first inequality in the second line of (3), does not give any information on the lower growth rate of f in terms of a power function with exponent $p > 1$.

A first investigation of variational problems with the structure (3) under some additional balancing conditions is due to Fuchs and Osmolovskii ([FO]), where Theorem 2 is shown in the case that $\mu < 4/n$.

Theorems of type 1 and 3 are established by Fuchs and Mingione (see [FuM]) – their assumptions on μ and q are discussed in Section 3.5.

C LINEAR GROWTH

It remains to discuss the case of variational problems with linear growth. On account of the lack of compactness in the non-reflexive Sobolev space $W_1^1(\Omega; \mathbb{R}^N)$, the problem $(\mathcal{P})$ in general fails to have solutions. Thus one either has to introduce a suitable notion of generalized minimizers (possibility $i)$) or one must pass to the dual variational problem (possibility $ii)$).

ad $i)$. Since the integrand f under consideration is of linear growth, any J-minimizing sequence $\{u_m\}$, $u_m \in u_0 + \overset{\circ}{W}_1^1(\Omega; \mathbb{R}^N)$, is uniformly bounded in the space $BV(\Omega; \mathbb{R}^N)$. This ensures the existence of a subsequence (not relabeled) and a function u in $BV(\Omega; \mathbb{R}^N)$ such that $u_m \to u$ in $L^1(\Omega; \mathbb{R}^N)$. Thus, one suitable definition of a generalized minimizer u is to require $u \in \mathcal{M}$, where the set $\mathcal{M}$ is given by

$$\mathcal{M} = \{u \in BV(\Omega; \mathbb{R}^N) : u \text{ is the } L^1\text{-limit of a } J\text{-minimizing sequence}$$

$$\text{from } u_0 + \overset{\circ}{W}_1^1(\Omega; \mathbb{R}^N)\} .$$

Another point of view is to define a relaxed functional $\hat{J}$ on the space $BV(\Omega; \mathbb{R}^N)$ (a precise notion of relaxation is given in Appendix A). Then generalized solutions of the problem $(\mathcal{P})$ are introduced as minimizers of a relaxed problem $(\hat{\mathcal{P}})$.

Remark 1.1. *We already like to mention that these formally different points of view in fact lead to the same set of functions. Moreover, the third approach to the definition of generalized minimizers given in [Se1], [ST] also leads to the same class of minimizing objects.*

ad $ii)$. Following [ET] we write

$$J[w] = \sup_{\tau \in L^\infty(\Omega; \mathbb{R}^{nN})} l(w, \tau) , \quad w \in u_0 + \overset{\circ}{W}_1^1(\Omega; \mathbb{R}^N) ,$$

where $l(w, \tau)$ denotes some natural Lagrangian (see Section 2.1.1). If we let

$$R : L^\infty(\Omega; \mathbb{R}^{nN}) \to \overline{\mathbb{R}} ,$$

$$R(\tau) := \inf_{u \in u_0 + \overset{\circ}{W}_1^1(\Omega; \mathbb{R}^N)} l(u, \tau) = \begin{cases} -\infty , & \text{if div } \tau \neq 0 , \\ l(u_0, \tau) , & \text{if div } \tau = 0 , \end{cases}$$

then the dual problem reads as

$$\text{to maximize } R \text{ among all functions in } L^\infty\left(\Omega; \mathbb{R}^{nN}\right), \qquad (\mathcal{P}^*)$$

where the existence of solutions easily is established.

In any of the above definitions the set of generalized minimizers of the problem $(\mathcal{P})$ may be very "large". In contrast to this fact, the solution of the dual problem is unique (see the discussion of Section 2.2). Moreover, the dual solution σ admits a clear physical or geometrical interpretation, for instance as a stress tensor or the normal to a surface. Hence, in the linear growth situation we wish to complete the above theorems by analogous regularity results for σ.

C.1 GEOMETRIC PROBLEMS OF LINEAR GROWTH

One of the most important (scalar) examples is the minimal surface case $f(Z) = \sqrt{1 + |Z|^2}$. A variety of references is available for the study of this variational integrand, let us mention the monographs of Giusti ([Giu2]) and Giaquinta/Modica/Souček ([GMS2]) at this point.

At first sight, ellipticity now is very bad since the inequalities in the second line of (3) just hold for the choices $\mu = 3$ and $q = 1$. On the other hand, this rough estimate is not needed because it is possible to benefit from the geometric structure of the problem (see Remark 4.3). A class of integrands with this structure is studied, for instance, in [GMS1] following the a priori gradient bounds given in [LU2]. It turns out that in the minimal surface case generalized $\hat{J}$-minimizers are of class $C^{1,\alpha}(\Omega)$ and that we have uniqueness up to a constant.

C.2 LINEAR GROWTH PROBLEMS WITHOUT GEOMETRIC STRUCTURE

The theory of perfect plasticity provides another famous variational integrand of linear growth. In this case the assumptions of smoothness and strict convexity imposed on f are no longer satisfied. Nevertheless, the example should be included in our discussion since we will benefit in Chapter 2 from the studies of Seregin ([Se1]–[Se6]) on this topic (compare the recent monograph [FuS2]).

The quantity of physical interest is the stress tensor σ, which is only known to be partially regular (compare [Se4]). Even in the two-dimensional setting $n = 2$ we just have some additional information on the singular set (see [Se6]) and the model of plastic materials with logarithmic hardening (as described in B.1) serves as a regular approximation.

It is already mentioned above that the vector-valued linear growth situation is covered by [AG2], provided that we restrict ourselves to smooth and strictly convex integrands. Anzellotti and Giaquinta prove Theorem 2 for generalized $\hat{J}$-minimizers, hence the same regularity result turns out to be true for any $u \in \mathcal{M}$ (see Section 2.3.1 for details). It remains to study the properties of the dual solution which (as noted above) for linear growth problems is a quantity of particular interest.

Before we summarize this brief overview in the table given below, we like to mention that of course there is a variety of further contributions where the class of admissible energy densities is equipped with some additional structure (see [AF4], [Lie2], [UU] and many others).

SOME KNOWN REGULARITY RESULTS IN THE CONVEX CASE

	$N = 1$	$N > 1$
A.1	(1) DeGiorgi, Moser, Nash, 　　Ladyzhenskaya/Ural'tseva 　　$\leq$ '65	(2) Anzellotti/Giaquinta '88 (3) Uhlenbeck, '77
A.2	(1) $1 < p \leq q < \dots$ 　　Marcellini $\approx$ '90	(2) $2 \leq p \leq q < \dots$ 　　Acerbi/Fusco '94, (3) bounded $\dots$, Choe '92
B.1	see $N > 1$	(3) $n = 2$: Frehse/Seregin '98 (2) $n \leq 4$: Fuchs/Seregin '98 (2) Esposito/Mingione '00 (3) Mingione/Siepe '99
B.2	(1) $\mu < 1 + 2/n$, $q < \dots$ 　　Fuchs/Mingione '00	(2) $\mu \leq 4/n$, "balanced" 　　Fuchs/Osmolovskii '98 (3) [FM] (see $N = 1$)
C.1	(1)$_{\hat{\jmath}}$ Giaquinta/Modica/ 　　Souček '79	—
C.2	—	(2)$_{\hat{\jmath}}$　[AG] (see A.1, $N > 1$) (P)$_{\sigma,\text{pl}}$ Seregin $\approx$ '90

(1), (2), (3): Theorems 1–3, respectively

(1)$_{\hat{\jmath}}$, (2)$_{\hat{\jmath}}$: corresponding results for generalized $\hat{J}$-minimizers

(P)$_{\sigma,\text{pl}}$: partial regularity for the stress tensor in the theory of perfect plasticity

In the following we are going to

- have a close look at linear growth problems;
- unify the results of A and B by the way including new classes of integrands;
- discuss the substantial extensions which follow in cases A, B and C from a natural boundedness condition.

Our main line skips from linear to superlinear growth and vice versa: in spite of the essential differences, these two items are strongly related by a non-uniform ellipticity condition (see Definition 3.4 and Assumption 4.1), by the applied techniques and to a certain extent by the obtained results. In particular, this relationship becomes evident while studying scalar variational problems with

- mixed anisotropic linear/superlinear growth conditions.

As the first center of interest, the discussion starts in Chapter 2 by considering the general linear growth situation. Here no uniqueness results for generalized minimizers can be expected and we concentrate on the dual solution σ which, according to the above remarks, is a reasonable physical point of view. The main contributions are

i) uniqueness of the dual solution under very weak assumptions;

ii) partial $C^{1,\alpha}$-regularity for weak cluster points of J-minimizing sequences and, as a consequence, partial $C^{0,\alpha}$-regularity for σ;

iii) a proof of the duality relation $\sigma = \nabla f(\nabla^a u^*)$ for a class of degenerate variational problems with linear growth. Here $\nabla^a u^*$ denotes the absolutely continuous part of ∇u^* with respect to the Lebesgue measure.

ad *i*). Standard arguments from convex analysis (compare [ET]) yield the uniqueness of the dual solution by assuming the conjugate function f^* to be strictly convex. We do not want to impose this condition since it is formulated in terms of f^*, hence there might be no easy way to check this assumption. In fact, using more or less elementary arguments, it is proved in Section 2.2 that there is no need to involve the conjugate function in an uniqueness theorem for the dual solution (see [Bi1]).

ad *ii*). Following the lines of [GMS1], any weak cluster point $u \in \mathcal{M}$ minimizes the relaxed problem $(\widehat{\mathcal{P}})$ associated to the original problem (see Appendix A.1). Alternatively (and as outlined in [BF1]), a local approach is preferred in Section 2.3.1 (see Remark 2.16 for a brief comment). In any case, the results of Anzellotti and Giaquinta apply and u is seen to be of class $C^{1,\alpha}$ on the non-degenerate regular set Ω_u (see (23), Section 2.3). As a next step, the duality relation $\sigma = \nabla f(\nabla u^*)$, $x \in \Omega_{u^*}$, is shown for a particular solution u^*, hence σ is of class $C^{0,\alpha}$ on this set.

ad *iii*). The duality relation is proved using local $C^{1,\alpha}$-results for some u^* as above. As a consequence, information on the behavior of σ is only obtained on the u^*-regular set. In Section 2.4, the almost everywhere identity

$\sigma = \nabla f(\nabla^a u^*)$ is established for a class of degenerate problems which gives intrinsic regularity results in terms of σ (this is due to [Bi2]). Note that the applied technique completely differs from the previous considerations since we cannot rely on regularity results: arguments from measure theory are combined with the construction of local comparison functions (see Appendix B.3).

Chapter 3 deals with the nearly linear and/or anisotropic situation. Here

   i) we introduce the notion of integrands with (s, μ, q)-growth;

and give a unified and extended approach to

  ii) the results of type (1) and (3) outlined in the above table;
 iii) the corresponding theorems (2).

Finally, reducing the generality of the previous sections, a theorem on

  iv) full $C^{1,\alpha}$-regularity of solutions of two-dimensional vector-valued problems with anisotropic power growth

completes Chapter 3.

ad i). The main observation is clarified in Example 3.7. Three free parameters occurring in the structure and growth conditions imposed on the integrand f determine the behavior of solutions, which now uniquely exist in an appropriate energy class: the growth rate s of the integrand f under consideration, and the exponents μ, q of a non-uniform ellipticity condition. This leads to the notion of integrands with (s, μ, q)-growth which includes and extends the list given in A and B in a natural way. Note that related structure conditions for variational integrands with superquadratic growth are introduced in [Ma5]–[Ma7] (see Section 3.5 for a brief discussion).

ad ii). Since regular solutions cannot be expected for the whole range of s, μ and q (we already mentioned [Gia2]), we impose the so called (s, μ, q)-condition. Observe that we do not lose information in comparison with the known results (see Section 3.5).

As a next step, uniform a priori L^q_{loc}-estimates for the gradients of a regularizing sequence are proved. This enables us to apply DeGiorgi-type arguments with uniform local a priori gradient bounds as the result. The conclusion then follows in a well known manner (we refer to [BFM] for a discussion of scalar variational problems with (s, μ, q)-growth).

It should be emphasized that the proof covers the whole scale of (s, μ, q)-integrands without distinguishing several cases.

ad iii). Here a blow-up procedure (compare [Ev], [CFM]) is used to prove partial regularity in the above setting (compare [BF2]). This generalizes the known results to a large extent (see Section 3.5).

ad *iv*). With the higher integrability results of the previous sections it is possible (following [BF6]) to refer to a lemma due to Frehse and Seregin.

In Chapter 4 we return to problems with linear growth, where we first benefit from some of the techniques outlined in Chapter 3, i.e.

i) a regular class of μ-elliptic integrands with linear growth is introduced.

Then the results are substantially improved by

ii) studying bounded solutions (in some natural sense);
iii) considering two-dimensional problems.

We finish the study of linear growth problems by proving the

iv) sharpness of the results.

ad *i*). Example 3.9 also provides a class of μ-elliptic integrands with linear growth in the sense that for all $Z, Y \in \mathbb{R}^{nN}$

$$\lambda \left(1 + |Z|^2\right)^{-\frac{\mu}{2}} |Y|^2 \leq D^2 f(Z)(Y,Y) \leq \Lambda \left(1 + |Z|^2\right)^{-\frac{1}{2}} |Y|^2 \tag{4}$$

holds for some $\mu > 1$ and with constants λ, Λ. If $\mu < 1 + 2/n$, then this class is called a regular one since generalized minimizers are unique up to a constant and since Theorems 1 and 3 for functions $u \in \mathcal{M}$ will be established following the arguments of Chapter 3 (see [BF3]). Let us shortly discuss the limitation $\mu < 1 + 2/n$. Given a suitable regularization u_δ, it is shown that

$$\omega_\delta := \left(1 + |\nabla u_\delta|^2\right)^{\frac{2-\mu}{4}}$$

is uniformly bounded in the class $W^1_{2,loc}(\Omega)$. This provides no information at all if the exponent is negative, i.e. if $\mu > 2$. An application of Sobolev's inequality, which needs the bound $\mu < 1 + 2/n$, proves uniform local higher integrability of the gradients. The final DeGiorgi-type arguments will lead to the same limitation on the ellipticity exponent μ.

ad *ii*). The minimal surface integrand can be interpreted as a μ-elliptic example with limit exponent $\mu = 3$ (recall that in the minimal surface case the regularity of solutions is obtained by using the geometric structure).

Section 4.2 and [Bi4] are devoted to the question, whether the limit $\mu = 3$ is of some relevance if the geometric structure condition is dropped. To this purpose some examples are discussed.

Then, imposing a natural boundedness condition, we prove even in the vector-valued setting (without assuming $f(Z) = g(|Z|^2)$) that a generalized minimizer u^* of class $W^1_1(\Omega; \mathbb{R}^N)$ exists. Moreover, u^* uniquely (up to a constant) determines the solutions of the problem

$$\int_\Omega f(\nabla w)\, dx + \int_{\partial \Omega} f_\infty\left((u_0 - w) \otimes \nu\right) d\mathcal{H}^{n-1} \to \min \quad \text{in } W^1_1\left(\Omega; \mathbb{R}^N\right). \tag{$\mathcal{P}'$}$$

If, as a substitute for the geometric structure, $\mu < 3$ is assumed, then the uniqueness of generalized minimizers up to a constant as well as Theorem 1 and Theorem 3 are true.

As indicated above, the proof of i) does not extend to these results: in Sections 4.2.2.1 and 4.2.3 we do not differentiate the Euler equation, thus we avoid to use Sobolev's inequality. Moreover, in the case $\mu < 3$, a preliminary iteration gives uniform L_{loc}^p-gradient bounds for any p. This is the reason why we may use Hölder's inequality and finally adjust the DeGiorgi iteration exponent to get the conclusion.

ad iii). It turns out (compare [Bi5]) that a boundedness condition is superfluous to establish the results of ii) in the two-dimensional case $n = 2$ (with the usual structure in the vector-valued setting). Note that, once more, $\mu = 3$ is exactly the limit case within reach.

ad iv). Extending the ideas of [GMS1], an example is given which shows that the problem $(\mathcal{P}')$ in general does not admit a W_1^1-solution if the ellipticity condition merely holds for some $\mu > 3$. Since the energy density under consideration is explicitely depending on x, we have to show first in Section 4.2.2.2 (as a model case) that a smooth x-dependence does not affect the above mentioned theorems, thus our example really is a counterexample (see also [BF8]).

Chapter 5 once more deals with the study of superlinear growth problems, where a boundedness condition analogous to Chapter 4.2 is supposed to be valid. We prove (in addition referring to [BF7], [BF9])

i) higher integrability and, as a corollary, a theorem of type (2) for variational integrands with a wide range of anisotropy.

Then, as a model case,

ii) scalar obstacle problems are studied for this class of energy densities and we prove a theorem of type (1).

ad i). Recalling the ideas of Chapter 4 we expect that these techniques may be applied to improve the results of Section 3.3 and Section 3.4 for bounded solutions in the case of variational integrals with superlinear growth. If we consider integrands with anisotropic (p, q)-growth, then the corresponding relation between p and q should read as $q < p+2$. However, as proved in [BF5], the "linear growth techniques" just yield $W_{q,loc}^1$-solutions if $q < p + 2/3$. The reason for this "lack of anisotropy" is the following: in Section 4.2 we could benefit from the growth rate $1 = q$ of the main quantity $\nabla f(Z) : Z$ under consideration. In the anisotropic superlinear case however, we just have the lower bound $p < q$ of this quantity. This is the reason why we change methods again and give a refined study of an Ansatz which traces back to [Ch]. As a result, the full correspondence to the linear growth situation is established, i.e. with the assumption

$$\lambda \left(1+|Z|^2\right)^{-\frac{\mu}{2}} |Y|^2 \leq D^2 f(Z)(Y,Y) \leq \Lambda \left(1+|Z|^2\right)^{\frac{q-2}{2}} |Y|^2$$

for all $Z, Y \in \mathbb{R}^{nN}$ with positive constants λ, Λ and for exponents $\mu < 3$, $q > 1$, higher local integrability follows from $q < 4-\mu$. This provides (together with some natural hypothesis) a corollary on partial regularity.

ad ii). Here, as a model case, we include the study of scalar obstacle problems. The methods as described in i) yield full $C^{1,\alpha}_{loc}$-regularity under the same condition $q < 4 - \mu$ which is quite weak (recall the counterexample of Section 4.4).

Chapter 6 (see [Bi6]) closes the line with the consideration of

- scalar variational problems with mixed anisotropic linear/superlinear growth conditions.

Here, on one hand, we essentially have to rely on the wide range of anisotropy which is admissible on account of Chapter 5. On the other hand, a refined study of the dual problem is needed since a dual solution may even fail to exist. This is caused by a possible anisotropic behavior of the superlinear part itself. Nevertheless, we obtain locally regular and uniquely determined (up to a constant) generalized minimizers which in return provide a "local stress tensor".

We finish our studies with three appendices:

the first one identifies the different ways to define generalized minimizers (recall Remark 1.1). The main Theorem A.6 (see [BF4]) proves, as a corollary, the uniqueness results applied in Chapter 4 which are based on the different approaches, respectively.

In Appendix B some density results are collected, where either a rigorous proof is hardly found in the literature or the claims have to be adjusted to the situation at hand. Maybe, the construction of local comparison functions given in Section B.3 is the only result which is unknown to the reader (compare [BF1]). This helpful lemma is used several times studying linear growth problems.

It is outlined in Appendix C (see [BF10], [ABF]) that the methods discussed throughout this monograph at least partially extend to the study of generalized Newtonian fluids. We did not include this material in the previous sections in order to keep the main line of the standard setting of the calculus of variations.

Variational problems with linear growth: the general setting

Following the main line sketched in the introduction, we start by considering the general linear growth situation. Recall that the variational problem $(\mathcal{P})$ then may fail to have solutions which leads to suitable notions of generalized minimizers. On the other hand, it is quite natural to introduce the dual variational problem which is of particular interest in the setting of this chapter since no uniqueness results on generalized minimizers are available.

Thus, we first have to give some introductory remarks on convex analysis in Section 2.1.1 in order to obtain a precise definition of the dual variational problem $(\mathcal{P}^*)$.

A first analysis of the dual solution(s) is given in Section 2.1.2: here a regularizing sequence is constructed which, in Lemma 2.6, is shown to converge to a maximizer of $(\mathcal{P}^*)$.

As a first regularity result, we prove that this maximizer is of class $W^1_{2,loc}(\Omega; \mathbb{R}^{nN})$ (compare Section 2.1.3).

One essential motivation for the study of the dual variational problem is the uniqueness of solutions. In Section 2.2 such a uniqueness result for the dual solution σ is derived under very weak assumptions, in particular Theorem 2.15 does not depend on the strict convexity of the conjugate function.

Two theorems of type (2) are outlined in the next section: each L^1-cluster point u^* of a J-minimizing sequence solves some relaxed problem $(\hat{\mathcal{P}})$ (see Remark 2.16 and Appendix A.1) which, on account of [AG2], implies $C^{1,\alpha}$-regularity on the non-degenerate regular set Ω_{u^*}. We then establish the existence of u^* as above such that the duality relation $\sigma = \nabla f(\nabla u^*)$ holds almost everywhere on Ω_{u^*}. This yields $C^{0,\alpha}$-regularity of the dual solution σ on this set.

In Section 2.4 we have a more detailed look at the degenerate situation: the above results for σ are formulated in terms of Ω_{u^*}, i.e. they involve the regular set of a special generalized minimizer. In order to obtain an intrinsic theory, we now prove that the duality relation in fact holds almost everywhere for a certain class of degenerate problems.

Throughout this chapter the variational integrand is supposed to satisfy the following general hypothesis:

Assumption 2.1. *The function f is smooth, strictly convex and of linear growth in the following sense:*

i) $f \in C^2 (\mathbb{R}^{nN})$.

ii) $f((1 - \lambda)Z + \lambda Y) < (1 - \lambda) f(Z) + \lambda f(Y)$ *for all* $Z \neq Y \in \mathbb{R}^{nN}$ *and for all* $0 < \lambda < 1$. *Suppose further that there is a positive number* ν_1 *such that for all* $Z, Y \in \mathbb{R}^{nN}$

$$0 \leq D^2 f(Z)(Y,Y) \leq \nu_1 \frac{1}{\sqrt{1 + |Z|^2}} |Y|^2 .$$

iii) *There is a real number* $\nu_2 > 0$ *such that* $|\nabla f(Z)| \leq \nu_2$ *for all* $Z \in \mathbb{R}^{nN}$.

iv) *For numbers* $\nu_3 > 0$ *and* $\nu_4 \in \mathbb{R}$ *we have* $f(Z) \geq \nu_3 |Z| + \nu_4$ *for all* $Z \in \mathbb{R}^{nN}$.

For the sake of simplicity the boundary values u_0 under consideration are supposed to be of class $W_2^1 (\Omega; \mathbb{R}^N)$. As outlined in Remark 2.5, this restriction on the boundary data can easily be removed.

2.1 Construction of a solution for the dual problem which is of class $W_{2,loc}^1 (\Omega; \mathbb{R}^{nN})$

We are going to give some introductory remarks on the dual problem associated to $(\mathcal{P})$. Moreover, a suitable regularization is introduced in Section 2.1.2. As an immediate consequence we obtain in Section 2.1.3 a maximizer σ of the problem $(\mathcal{P}^*)$ which is of class $W_{2,loc}^1 (\Omega; \mathbb{R}^{nN})$.

2.1.1 The dual problem

Here we recall some well known facts from convex analysis leading to the notion of the dual problem. As a reference one may choose, for instance, [Ro] or [Ze], we mostly follow the book of Ekeland and Temam ([ET]).

Definition 2.2. *Consider a Banach space V, its dual V^* and a function $G : V \rightarrow \overline{\mathbb{R}}$. Then the polar or conjugate function of G is defined for all $v^* \in V^*$ by*

$$G^* (v^*) := \sup_{v \in V} \{ \langle v, v^* \rangle - G(v) \} .$$

The bipolar function is given for all $v \in V$ by

$$G^{**} (v) := \sup_{v^* \in V^*} \{ \langle v, v^* \rangle - G^* (v^*) \} .$$

Since we always consider lower semicontinuous and convex functions G, the bipolar function satisfies (see [ET], Prop. 4.1, p. 18)

$$G^{**}(v) = G(v) \quad \text{for all } v \in V . \tag{1}$$

If the subdifferential of G is denoted by ∂G (see [ET] pp. 20) and if $\partial G(v) \neq \emptyset$, then we have the duality relation:

$$v^* \in \partial G(v) \iff G(v) + G^*(v^*) = \langle v, v^* \rangle .$$

This gives for our smooth integrand $f : \mathbb{R}^{nN} \to \mathbb{R}$:

$$f(w) + f^*(\nabla f(w)) = w : \nabla f(w) . \tag{2}$$

Here and in what follows the symbol $Z : Y$ is used to denote the standard scalar product in $\mathbb{R}^{nN}$. We next derive an alternative expression for $J[w]$, $w \in W_1^1(\Omega; \mathbb{R}^N)$: given f as above, we consider the functional $G : L^1(\Omega; \mathbb{R}^{nN}) \to \overline{\mathbb{R}}$,

$$G(p) := \int_\Omega f(p) \, dx \quad \text{for all } p \in L^1(\Omega; \mathbb{R}^{nN}) .$$

Then Proposition 2.1, [ET], p. 271, can be applied, hence, together with (1) and the definition of the conjugate function, we see that

$$\int_\Omega f(p) \, dx = G(p) = G^{**}(p)$$

$$= \sup_{\varkappa \in L^\infty(\Omega; \mathbb{R}^{nN})} \left\{ \int_\Omega \varkappa : p \, dx - G^*(\varkappa) \right\}$$

$$= \sup_{\varkappa \in L^\infty(\Omega; \mathbb{R}^{nN})} \left\{ \int_\Omega \varkappa : p \, dx - \int_\Omega f^*(\varkappa) \, dx \right\} .$$

This formula holds for all $p \in L^1(\Omega; \mathbb{R}^{nN})$, in particular for $p = \nabla w$, $w \in W_1^1(\Omega; \mathbb{R}^N)$. We obtain the representation formula

$$J[w] = \sup_{\varkappa \in L^\infty(\Omega; \mathbb{R}^{nN})} \left\{ \int_\Omega \varkappa : \nabla w \, dx - \int_\Omega f^*(\varkappa) \, dx \right\} . \tag{3}$$

Remark 2.3. *Using the notation introduced by Ekeland and Temam, Chapter III.4, pp. 58, we arrive at (3) if we set $J[w] = G(w, \Lambda w)$ and $\Phi(w, p) = J(w, \Lambda w - p)$, where the linear operator Λ is the ∇-operator.*

The representation formula (3) motivates to define the Lagrangian $l(w, \varkappa)$ for all $(w, \varkappa) = (u_0 + \varphi, \varkappa)$ in the class $(u_0 + \overset{\circ}{W}_1^1(\Omega; \mathbb{R}^N)) \times L^\infty(\Omega; \mathbb{R}^{nN})$ by the formula

$$l(w, \varkappa) := \int_\Omega \varkappa : \nabla w \, dx - \int_\Omega f^*(\varkappa) \, dx = l(u_0, \varkappa) + \int_\Omega \varkappa : \nabla \varphi \, dx .$$

Now, the dual functional $R\colon L^\infty(\Omega; \mathbb{R}^{nN}) \to \overline{\mathbb{R}}$ is given by

$$R[\varkappa] := \inf_{w \in u_0 + \overset{\circ}{W}_1^1(\Omega;\mathbb{R}^N)} l(w, \varkappa) \,,$$

and the dual problem reads as:

$$\text{to maximize } R \text{ among all functions } \varkappa \in L^\infty(\Omega; \mathbb{R}^{nN}) \,. \qquad (\mathcal{P}^*)$$

Remark 2.4.

i) *The definition of R shows that we have for any $\varkappa \in L^\infty(\Omega; \mathbb{R}^{nN})$*

$$R[\varkappa] = \begin{cases} -\infty & \text{if } \operatorname{div} \varkappa \neq 0 \,, \\ l(u_0, \varkappa) & \text{if } \operatorname{div} \varkappa = 0 \,. \end{cases}$$

ii) *On the complement of the set $\overline{\operatorname{Im}(\nabla f)}$ we have the identity $f^* \equiv +\infty$, thus for any $\varkappa$ under consideration we may assume that $\varkappa(x) \in \overline{\operatorname{Im}(\nabla f)}$ almost everywhere.*

Let us finally mention one essential property of the dual functional:

$$\inf_{w \in u_0 + \overset{\circ}{W}_1^1(\Omega;\mathbb{R}^N)} J[w] = \sup_{\varkappa \in L^\infty(\Omega;\mathbb{R}^{nN})} R[\varkappa] \,. \qquad (4)$$

Here the estimate

$$J[w] = \sup_{\varkappa \in L^\infty(\Omega;\mathbb{R}^{nN})} l(w, \varkappa) \geq \sup_{\varkappa \in L^\infty(\Omega;\mathbb{R}^{nN})} \inf_{v \in u_0 + \overset{\circ}{W}_1^1(\Omega;\mathbb{R}^N)} l(v, \varkappa)$$

$$= \sup_{\varkappa \in L^\infty(\Omega;\mathbb{R}^{nN})} R[\varkappa]$$

is obvious, for the opposite inequality we either refer to [ET] or to the following Lemma 2.6, where the inf-sup relation is proved as a byproduct (compare Remark 2.9, i)).

2.1.2 Regularization

Approximating the original problem in a well known way (compare, e.g. [Se4]), a special maximizing sequence for the dual problem is constructed in this subsection. Here we have to recall that the boundary values u_0 are assumed to be of class $W_2^1(\Omega; \mathbb{R}^N)$.

The problem $(\mathcal{P})$ is approximated in the following way: for any $0 < \delta < 1$ we consider the functional

$$J_\delta[w] := \frac{\delta}{2} \int_\Omega |\nabla w|^2 \, dx + J[w] \,, \quad w \in u_0 + \overset{\circ}{W}_2^1(\Omega; \mathbb{R}^N) \,,$$

and denote by u_δ the unique solution of

$$J_\delta[w] \to \min , \quad w \in u_0 + \overset{\circ}{W}{}^1_2(\Omega; \mathbb{R}^N) . \qquad (\mathcal{P}_\delta)$$

Setting

$$f_\delta := \frac{\delta}{2}|\cdot|^2 + f , \quad \tau_\delta := \nabla f(\nabla u_\delta) , \quad \sigma_\delta := \delta \nabla u_\delta + \tau_\delta = \nabla f_\delta(\nabla u_\delta) ,$$

we have the Euler equation

$$\int_\Omega \sigma_\delta : \nabla \varphi \, dx = 0 \quad \text{for all } \varphi \in \overset{\circ}{W}{}^1_2(\Omega; \mathbb{R}^N) . \qquad (5)$$

The minimality of u_δ implies $J_\delta[u_\delta] \le J_\delta[u_0] \le J_1[u_0]$, hence there are positive constants c_1, c_2 such that

$$\delta \int_\Omega |\nabla u_\delta|^2 \, dx \le c_1 , \quad \int_\Omega f(\nabla u_\delta) \, dx \le c_2 . \qquad (6)$$

Remark 2.5. *If we consider boundary values of class $W^1_1(\Omega; \mathbb{R}^N)$, then the above regularization has to be applied to an approximating sequence $\{u_0^m\} \subset C^\infty(\overline{\Omega}; \mathbb{R}^N)$ of boundary values converging in $W^1_1(\Omega; \mathbb{R}^N)$ to u_0. As a result, the regularized sequence depends on m, i.e. $u_\delta = u_\delta^m$. It will turn out in this chapter that this is no difference at all provided we have the uniform a priori bound (6). This, however, can be achieved by choosing $\delta = \delta(m)$ sufficiently small. For details we refer to [Bi5].*

The first inequality of (6) immediately gives

$$\|\delta \nabla u_\delta\|^2_{L^2(\Omega; \mathbb{R}^{nN})} = \delta \left(\delta \int_\Omega |\nabla u_\delta|^2 \, dx \right) \to 0 \quad \text{as } \delta \to 0 . \qquad (7)$$

Since ∇f is bounded, we may also assume that

$$\|\tau_\delta\|_{L^\infty(\Omega; \mathbb{R}^{nN})} \le c .$$

Passing to a subsequence (which is not relabeled) we obtain limits $\tau \in L^\infty(\Omega; \mathbb{R}^{nN})$ and $\sigma \in L^2(\Omega; \mathbb{R}^{nN})$ such that $\tau_\delta \overset{*}{\rightharpoonup} \tau$ in $L^\infty(\Omega; \mathbb{R}^{nN})$ as well as

$$\sigma_\delta \rightharpoonup \sigma = \tau \quad \text{in } L^2(\Omega; \mathbb{R}^{nN}) \quad \text{as } \delta \to 0 .$$

Note that the convergence of a subsequence $\{\sigma_\delta\}$ yields by (7) the convergence of the corresponding subsequence $\{\tau_\delta\}$ and vice versa.

The following lemma shows that we have produced a maximizer of the dual variational problem.

Lemma 2.6.

i) *Any weak L^2-cluster point σ of the sequence $\{\sigma_\delta\}$ is admissible in the sense that we have div $\sigma = 0$.*

ii) Any weak L^2-cluster point σ of the sequence $\{\sigma_\delta\}$ maximizes the dual variational problem $(\mathcal{P}^*)$.

Remark 2.7. *Note that a strict convexity condition for the dual function f^* is not imposed, i.e. the uniqueness of maximizers remains to be proved (compare Section 2.2).*

Remark 2.8. *Lemma 2.6 corresponds to Lemma 2 of [Se4] where the case of integrands depending on the modulus of the gradient is considered. Similar results were obtained in [Se5], Lemma 3.2, and [Se6], Lemma 3.1.*

Proof of Lemma 2.6. Equation (5) yields for the sequence $\{\sigma_\delta\}$ under consideration

$$\int_\Omega \tau_\delta : \nabla\varphi \, dx + \int_\Omega \delta\nabla u_\delta : \nabla\varphi \, dx = 0 \quad \text{for all } \varphi \in C_0^\infty(\Omega, \mathbb{R}^N) \, .$$

This implies *i*) by the above stated convergences.

To complete the proof, observe that the duality relation (given in (2)) $\tau_\delta : \nabla u_\delta - f^*(\tau_\delta) = f(\nabla u_\delta)$ implies

$$
\begin{aligned}
J_\delta[u_\delta] &= \frac{\delta}{2}\int_\Omega |\nabla u_\delta|^2 \, dx + \int_\Omega \left(\tau_\delta : \nabla u_\delta - f^*(\tau_\delta)\right) dx \\
&= -\frac{\delta}{2}\int_\Omega |\nabla u_\delta|^2 \, dx + \int_\Omega \left(\sigma_\delta : \nabla u_\delta - f^*(\tau_\delta)\right) dx \\
&= -\frac{\delta}{2}\int_\Omega |\nabla u_\delta|^2 \, dx + \int_\Omega \left(\sigma_\delta : \nabla u_0 - f^*(\tau_\delta)\right) dx \, ,
\end{aligned}
$$

where we used (5) to get the last equation. We have

$$
\begin{aligned}
\sup &\left\{ R[\varkappa] : \varkappa \in L^\infty(\Omega; \mathbb{R}^{nN}) \right\} \\
&\leq \inf \left\{ J[w] : w \in u_0 + \overset{\circ}{W}{}_1^1(\Omega; \mathbb{R}^N) \right\} \leq J[u_\delta] \leq J_\delta[u_\delta] \\
&= -\frac{\delta}{2}\int_\Omega |\nabla u_\delta|^2 \, dx + \int_\Omega \left(\sigma_\delta : \nabla u_0 - f^*(\tau_\delta)\right) dx \qquad (8) \\
&= -\frac{\delta}{2}\int_\Omega |\nabla u_\delta|^2 \, dx + \int_\Omega \left(\tau_\delta : \nabla u_0 - f^*(\tau_\delta)\right) dx \\
&\quad + \delta \int_\Omega \nabla u_\delta : \nabla u_0 \, dx \, .
\end{aligned}
$$

Passing to the limit, using the upper semicontinuity of $-\int_\Omega f^*(\cdot)\, dx$ with respect to weak-* convergence and observing that the last integral on the right-hand side of (8) tends to 0 as $\delta \to 0$, we obtain

$$\delta \int_\Omega |\nabla u_\delta|^2 \, dx \to 0 \quad \text{as } \delta \to 0 \qquad (9)$$

as well as the maximality of σ. $\quad\square$

Remark 2.9.

  i) *Replacing the first inequality in (8) by a strict one, the inf-sup relation (4) follows by contradiction.*
  ii) *Inequality (8) proves in addition that $\{u_\delta\}$ is a J-minimizing sequence.*

2.1.3 $W^1_{2,loc}$-regularity for the dual problem

The first regularity result for the dual problem reads as

Theorem 2.10. *Let σ denote a weak L^2-cluster point of the sequence $\{\sigma_\delta\}$. Then we have*

$$\sigma \in W^1_{2,loc}(\Omega; \mathbb{R}^{nN}) \, .$$

Remark 2.11. *Again we benefit from arguments outlined in [Se2] and [Se4]–[Se6].*

Proof of Theorem 2.10. We fix a converging sequence $\{\sigma_\delta\}$. Using the standard difference quotient technique, it is easily seen that u_δ (recall the notation of Section 2.1.2) is of class $W^2_{2,loc}(\Omega; \mathbb{R}^N)$. Moreover, since $|D^2 f_\delta|$ is bounded, $\nabla f_\delta(\nabla u_\delta)$ is of class $W^1_{2,loc}(\Omega; \mathbb{R}^{nN})$ with partial derivatives (almost everywhere)

$$\partial_\gamma \big(\nabla f_\delta(\nabla u_\delta) \big) = D^2 f_\delta(\nabla u_\delta) \big(\partial_\gamma \nabla u_\delta, \cdot \big) \, , \quad \gamma = 1, \ldots, n \, .$$

Now, given $\varphi \in C_0^\infty(\Omega; \mathbb{R}^N)$, we take $\partial_\gamma \varphi$, $\gamma = 1, \ldots, n$, as an admissible choice in the Euler equation (5). An integration by parts implies together with the above remarks

$$\int_\Omega D^2 f_\delta(\nabla u_\delta) \big(\partial_\gamma \nabla u_\delta, \nabla \varphi \big) \, dx = 0 \quad \text{for all } \varphi \in C_0^\infty(\Omega; \mathbb{R}^N) \, . \tag{10}$$

Using standard approximation arguments, (10) is seen to be true for all $\varphi \in W^1_2(\Omega; \mathbb{R}^N)$ which are compactly supported in Ω. In particular, if some ball $B_r(x_0) \Subset \Omega$ is fixed, then $\varphi = \partial_\gamma u_\delta \, \eta^2$, $\eta \in C_0^\infty(\Omega)$, $0 \leq \eta \leq 1$, $\eta \equiv 1$ on $B_r(x_0)$, is admissible in (10) and we obtain

$$\begin{aligned} I_1 &:= \int_\Omega D^2 f_\delta(\nabla u_\delta) \big(\partial_\gamma \nabla u_\delta, \partial_\gamma \nabla u_\delta \big) \eta^2 \, dx \\ &= -2 \int_\Omega D^2 f_\delta(\nabla u_\delta) \big(\partial_\gamma \nabla u_\delta, \partial_\gamma u_\delta \otimes \nabla \eta \big) \eta \, dx =: I_2 \, . \end{aligned} \tag{11}$$

Here we always take the sum with respect to $\gamma = 1, \ldots, n$. An upper bound for $|I_2|$ is given by (compare Assumption 2.1 and (6))

$$|I_2| \leq c I_1^{\frac{1}{2}} \left(\int_\Omega D^2 f_\delta(\nabla u_\delta)(\partial_\gamma u_\delta \otimes \nabla \eta, \partial_\gamma u_\delta \otimes \nabla \eta) \, dx \right)^{\frac{1}{2}}$$

$$\leq c(\nabla \eta) I_1^{\frac{1}{2}} \left(\int_\Omega \left[(1 + |\nabla u_\delta|^2)^{-\frac{1}{2}} + \delta \right] |\nabla u_\delta|^2 \, dx \right)^{\frac{1}{2}} \qquad (12)$$

$$\leq c I_1^{\frac{1}{2}} .$$

Now, the Cauchy-Schwarz inequality shows that we have almost everywhere

$$|\nabla \sigma_\delta|^2 = D^2 f_\delta(\nabla u_\delta)(\partial_\gamma \nabla u_\delta, \partial_\gamma \sigma_\delta)$$

$$\leq \left[D^2 f_\delta(\nabla u_\delta)(\partial_\gamma \nabla u_\delta, \partial_\gamma \nabla u_\delta) \right]^{\frac{1}{2}} \left[D^2 f_\delta(\nabla u_\delta)(\partial_\gamma \sigma_\delta, \partial_\gamma \sigma_\delta) \right]^{\frac{1}{2}} .$$

Since $|D^2 f_\delta|$ is uniformly bounded, there exists a constant, independent of δ, such that

$$|\nabla \sigma_\delta|^2 \leq c D^2 f_\delta(\nabla u_\delta)(\partial_\gamma \nabla u_\delta, \partial_\gamma \nabla u_\delta) . \qquad (13)$$

Combining (11)–(13) we have proved that the sequence $\{\sigma_\delta\}$ is uniformly (with respect to δ) bounded in $W^1_{2,loc}(\Omega; \mathbb{R}^{nN})$. This, together with the weak convergence of σ_δ, yields the theorem. $\square$

2.2 A uniqueness theorem for the dual problem

We now concentrate on the uniqueness of the dual solution which usually is established by assuming the conjugate function f^* to be strictly convex (see [ET], (3.34), p. 146). This hypothesis is formulated in terms of the conjugate function, hence, it might be difficult to verify the assumption for a given class of integrands f. Here we show by means of more or less elementary arguments that the strict convexity, the smoothness and the linear growth of f in the sense of Assumption 2.1 are sufficient to imply the uniqueness of the dual solution (without additional restrictions). The main idea is to construct (using Theorem 2.10) one special maximizer σ of the dual problem which is almost everywhere seen to be a mapping into the open set $\text{Im}(\nabla f)$. On this set f^* is known to be strictly convex by strict convexity of f. Thus, there is one solution σ for which we do not have to care about the fact that f^* on the closure of $\text{Im}(\nabla f)$ might not be strictly convex. This will give our uniqueness result.

Remark 2.12. *Alternatively, Corollary 2.18 together with the duality relation (31) of Section 2.3.2 could be used to provide one maximizer with values in the open set $\text{Im}(\nabla f)$. However, it should be emphasized that Theorem 2.15 also covers the degenerate case in the sense that $D^2 f(Z)(Y,Y) = 0$ is not excluded.*

Theorem 2.13 below is the main tool to prove the uniqueness of the dual solution. Here and in the following we let $U := \text{Im}(\nabla f)$.

Theorem 2.13. *Any weak L^2-cluster point σ of the δ-regularization given in Section 2.1.2 satisfies*

$$\left|\{x \in \Omega : \sigma(x) \in \partial U\}\right| = 0 \,.$$

Here $|\cdot|$ denotes the Lebesgue measure $\mathcal{L}^n$.

For the proof we need the following observation.

Lemma 2.14. *For all real numbers $K > 0$ there is an $\varepsilon > 0$ such that for all $Z \in \mathbb{R}^{nN}$*

$$\text{dist}\big(\nabla f(Z), \partial U\big) < \varepsilon \;\Rightarrow\; |Z| > K \,.$$

Proof of Lemma 2.14. Note that, by the strict convexity of f, $Z \neq Y \in \mathbb{R}^{nN}$ implies

$$\big[\nabla f(Z) - \nabla f(Y)\big] : (Z - Y) > 0 \,,$$

i.e. ∇f is a one-to-one mapping. In fact, we have

$$\big[\nabla f(Z) - \nabla f(Y)\big] : (Z - Y)$$
$$= \int_0^1 D^2 f\big(sZ + (1-s)Y\big)\big((Z-Y),(Z-Y)\big) \, ds \geq 0 \,.$$

Setting $g(s) := f\big(sZ+(1-s)Y\big)$, equality would give $g''(s) \equiv 0$ for all $s \in (0,1)$ which contradicts the strict convexity.

Now fix a real number $K > 0$. Since ∇f is continuous and one-to-one we may apply the Theorem on Domain Invariance (compare [Sch], Corollary 3.22, p. 77) to see that U is an open set. Thus

$$\nabla f\left(\overline{B_K(0)}\right) \Subset U$$

and there is an $\varepsilon = \varepsilon(K)$ such that

$$\text{dist}\left(\nabla f\left(\overline{B_K(0)}\right), \partial U\right) > \varepsilon \,.$$

This proves the lemma. $\quad\square$

Proof of Theorem 2.13. With the notation of Section 2.1.2 we consider a sequence $\delta_m \to 0$ as $m \to \infty$ such that the weak L^2-limit σ of $\{\sigma_{\delta_m}\}$ exists. Suppose by contradiction that there is a real number $\gamma > 0$ and a set $\Sigma \subset \Omega$ satisfying

$$|\Sigma| > \gamma \quad \text{and} \quad \sigma(x) \in \partial U \text{ for all } x \in \Sigma \,.$$

Here and in the following, sets of Lebesgue measure zero are neglected. If necessary, we always choose suitable subsequences from our given sequence $\{\delta_m\}$ – omitting further indices – such that all the limits below are well defined.

Since the approximating sequence σ_δ is uniformly bounded in the Sobolev class $W^1_{2,loc}(\Omega, \mathbb{R}^{nN})$ (compare Theorem 2.10) we may assume that

$$\sigma_\delta(x) \to \sigma(x) \quad \text{for all } x \in \Sigma \text{ as } \delta \to 0 .$$

Because of (7), Section 2.1.2, in addition

$$\delta \nabla u_\delta(x) \to 0 \quad \text{for all } x \in \Sigma \text{ as } \delta \to 0$$

can be assumed. Moreover, Egoroff's theorem yields a measurable set $E \subset \Omega$ such that

$$|\Omega - E| < \frac{\gamma}{2}$$

and such that

$$\sigma_\delta \rightrightarrows \sigma \quad \text{as well as} \quad \delta \nabla u_\delta \rightrightarrows 0 \quad \text{on } E \text{ as } \delta \to 0 . \tag{14}$$

The set E is measurable and

$$|\Sigma| = |\Sigma \cap E| + |\Sigma - E|$$

immediately implies

$$|\Sigma \cap E| > \frac{\gamma}{2} . \tag{15}$$

By construction, $\{u_\delta\}$ is a J-minimizing sequence (see Remark 2.9, ii)). In particular, on account of the linear growth of f, there is a real number $c_0 > 0$ such that for all δ sufficiently small

$$\int_\Omega |\nabla u_\delta| \, dx < c_0 . \tag{16}$$

Given c_0 we choose $K > 2c_0/\gamma$ and $0 < \varepsilon = \varepsilon(K)$ as determined in Lemma 2.14. The uniform convergence (14) shows that for all $x \in E$ and for all δ sufficiently small

$$|\sigma_\delta(x) - \sigma(x)| < \frac{\varepsilon}{2} \quad \text{and} \quad |\delta \nabla u_\delta(x)| < \frac{\varepsilon}{2} ,$$

hence, the definition of σ_δ gives

$$\left| \nabla f(\nabla u_\delta(x)) - \sigma(x) \right| < \varepsilon \quad \text{for all } x \in E .$$

If $x \in \Sigma \cap E$, then $\sigma(x) \in \partial U$ implies

$$\text{dist}\left(\nabla f(\nabla u_\delta(x)), \partial U \right) < \varepsilon .$$

Since $\varepsilon = \varepsilon(K)$ was chosen according to Lemma 2.14 we arrive at

$$|\nabla u_\delta(x)| > K \quad \text{for all } x \in \Sigma \cap E ,$$

and finally get, using (15), a contradiction to (16):

$$\int_\Omega |\nabla u_\delta(x)| \, dx \geq \int_{\Sigma \cap E} |\nabla u_\delta(x)| \, dx > \int_{\Sigma \cap E} K \, dx$$
$$> \frac{\gamma}{2} K > c_0 \, .$$

Thus, the theorem is proved. □

Theorem 2.15. *With the above Assumption 2.1 on f the dual problem $(\mathcal{P}^*)$ admits a unique solution σ.*

Proof. We fix a weak L^2-limit σ of the δ-approximation. By Lemma 2.6 this limit is known to be a solution of $(\mathcal{P}^*)$. Suppose by contradiction that the dual problem admits a second maximizer $\tilde{\sigma} \neq \sigma$. Now, Ω is divided into four parts,

$$\Omega = \Sigma_0 \cup \Sigma_1 \cup \Sigma_2 \cup \Sigma_3 \, ,$$

where we have by definition

$$\sigma(x) = \tilde{\sigma}(x) \qquad\qquad \text{for } x \in \Sigma_0 \, ,$$

$$\sigma(x) \neq \tilde{\sigma}(x) \, , \, \sigma(x) \in U \, , \, \tilde{\sigma}(x) \in U \quad \text{for } x \in \Sigma_1 \, ,$$

$$\sigma(x) \neq \tilde{\sigma}(x) \, , \, \sigma(x) \in U \, , \, \tilde{\sigma}(x) \in \partial U \quad \text{for } x \in \Sigma_2 \, ,$$

$$|\Sigma_3| = 0 \, .$$

The last equality is justified by Theorem 2.13 and by Remark 2.4, *ii*). The conjugate function f^* of f is a lower semicontinuous convex function on $\overline{U}$ and by (2) of Section 2.1.1 it is also seen to be strictly convex on U. This implies

$$|\Sigma_1| = 0 \, . \tag{17}$$

In fact, setting

$$\varkappa(x) := \frac{\sigma(x) + \tilde{\sigma}(x)}{2} \, ,$$

the convexity of f^* gives for almost every $x \in \Omega$

$$f^*(\varkappa(x)) \leq \frac{1}{2} f^*(\sigma(x)) + \frac{1}{2} f^*(\tilde{\sigma}(x)) \, ,$$

and the strict inequality holds for all $x \in \Sigma_1$. Recalling Remark 2.4, *i*), we obtain

$$R[\varkappa] = l(u_0, \varkappa)$$

$$= \int_\Omega \left(\frac{\sigma + \tilde{\sigma}}{2}\right) : \nabla u_0 \, dx - \int_{\Omega - \Sigma_1} f^*(\varkappa) \, dx - \int_{\Sigma_1} f^*(\varkappa) \, dx \tag{18}$$

$$\geq \int_\Omega \left(\frac{\sigma + \tilde{\sigma}}{2}\right) : \nabla u_0 \, dx - \frac{1}{2} \int_\Omega f^*(\sigma) \, dx - \frac{1}{2} \int_\Omega f^*(\tilde{\sigma}) \, dx$$

$$= \frac{1}{2} R[\sigma] + \frac{1}{2} R[\tilde{\sigma}] = \sup R \, .$$

Since the strict inequality would hold for $|\Sigma_1| > 0$, assertion (17) is proved by contradiction. We next claim that

$$|\Sigma_2| = 0 \, . \tag{19}$$

To prove (19), we fix $x \in \Sigma_2$, i.e.

$$P := \sigma(x) \in U \quad \text{and} \quad Q := \tilde{\sigma}(x) \in \partial U \, .$$

Now, U is an open set and Theorem 6.1, [Ro], p. 45, proves

$$[P, Q] := \{(1 - t)P + tQ : 0 \leq t < 1\} \subset U \, . \tag{20}$$

If we set for $m \in \mathbb{N}$ sufficiently large

$$Q_m := P + \frac{m-1}{m}(Q - P) \, ,$$

$$R_m := P + \frac{1}{2}(Q_m - P) = P + \frac{m-1}{2m}(Q - P) \, ,$$

and if we introduce for $t \in (0, 1)$ the function

$$g(t) = f^*(P + t(Q - P))$$

which, by (20), is known to be strictly convex, then we see (compare, for instance, the proof of Theorem 4.4, [Ro], p. 26) that there is a real number $0 < \varepsilon = \varepsilon(P, Q)$ such that

$$f^*(R_m) - f^*(P) = \int_0^{\frac{m-1}{2m}} g'(s) \, ds \leq g'\left(\frac{m-1}{2m}\right) \frac{m-1}{2m} - \varepsilon \, ,$$

$$f^*(Q_m) - f^*(R_m) = \int_{\frac{m-1}{2m}}^{\frac{m-1}{m}} g'(s) \, ds \geq g'\left(\frac{m-1}{2m}\right) \frac{m-1}{2m} + \varepsilon \, .$$

Combining these inequalities it is proved that

$$f^*(R_m) \leq \frac{1}{2} f^*(P) + \frac{1}{2} f^*(Q_m) - \varepsilon \, . \tag{21}$$

Since a lower semicontinuous convex function is continuous on straight lines to the boundary and since ε is not depending on m, we may pass in (21) to the limit $m \to \infty$ and obtain for all $x \in \Sigma_2$

$$f^*\big(\varkappa(x)\big) = f^*\left(\frac{\sigma(x) + \tilde{\sigma}(x)}{2}\right) < \frac{1}{2} f^*(\sigma(x)) + \frac{1}{2} f^*(\tilde{\sigma}(x)) \ .$$

The same computations as outlined in (18) give (19), we have shown

$$\Omega = \Sigma_0 \cup (\Omega - \Sigma_0) \quad \text{with} \quad |\Omega - \Sigma_0| = 0 \ ,$$

and the theorem is proved by the definition of Σ_0. $\square$

2.3 Partial $C^{1,\alpha}$- and $C^{0,\alpha}$-regularity, respectively, for generalized minimizers and for the dual solution

In the general setting of vector-valued variational problems with linear growth full regularity cannot be expected – even if we additionally assume that $D^2 f(Z) > 0$ holds for any matrix $Z \in \mathbb{R}^{nN}$. We prove in Section 2.3.1 that $C^{1,\alpha}$-regularity of generalized minimizers

$$u^* \in \mathcal{M} = \big\{ u \in BV\big(\Omega; \mathbb{R}^N\big) : u \text{ is the } L^1\text{-limit of a } J\text{-minimizing}$$

$$\text{sequence from } u_0 + \overset{\circ}{W}{}^1_1(\Omega; \mathbb{R}^N)\big\}$$

holds on the non-degenerate regular set Ω_{u^*}. To give a precise definition of this set, we observe that for almost all $x \in \Omega$ there exists a matrix $P \in \mathbb{R}^{nN}$ such that

$$\lim_{r \to 0} \frac{1}{|B_r(x)|} \int_{B_r(x)} |\nabla u^* - P| := \lim_{r \to 0} \frac{1}{|B_r(x)|} \int_{B_r(x)} |\nabla^a u^* - P| \, dx$$

$$+ \lim_{r \to 0} \frac{1}{|B_r(x)|} \int_{B_r(x)} |\nabla^s u^*| = 0 \ . \ (22)$$

Here $\nabla^a u^*$ denotes the absolutely continuous part of ∇u^* with respect to the Lebesgue measure, whereas $\nabla^s u^*$ is used as the symbol for the singular part. Then, the non-degenerate regular set is defined for $u^* \in \mathcal{M}$ via

$$\Omega_{u^*} := \big\{ x \in \Omega : (22) \text{ holds with } D^2 f(P) > 0 \big\} \ . \tag{23}$$

The main Theorem to study the smoothness of generalized minimizers is given in [AG2] where local minimizers of some relaxed functional are considered. In order to apply this theorem to $u^* \in \mathcal{M}$ as given above we like to remark

Remark 2.16. *One possibility is to follow the lines of [GMS1] and to work in the space* $BV_{u_0}(\Omega; \mathbb{R}^N)$. *This is outlined in Appendix A.1.*

We prefer a local approach which is based on the construction of suitable comparison functions as given in Appendix B.3. It will turn out in addition that this local point of view provides a helpful tool for the consideration of degenerate problems (compare Section 2.4). Another application of Lemma B.5 is found in Section 4.3.

In Section 2.3.2 it remains to prove the existence of $u^* \in \mathcal{M}$ such that $\sigma = \nabla f(\nabla u^*)$ holds almost everywhere on Ω_{u^*}. This gives the $C^{0,\alpha}$-regularity of the dual solution on this set. In particular, if we consider non-degenerate problems then the stress tensor is Hölder continuous on an open set of full measure.

2.3.1 Partial $C^{1,\alpha}$-regularity of generalized minimizers

Theorem 2.17. *Suppose that the integrand f satisfies the general Assumption 2.1. Moreover, consider a J-minimizing sequence* $\{u_m\}$ *from the affine class* $u_0 + \overset{\circ}{W}{}^1_1(\Omega; \mathbb{R}^N)$ *and* $u^* \in L^1(\Omega; \mathbb{R}^N)$ *satisfying*

$$u_m \to u^* \quad \text{in } L^1(\Omega; \mathbb{R}^N) \quad \text{as } m \to \infty .$$

If Ω_{u^*} *is given according to (23) then* Ω_{u^*} *is an open set and we have*

$$u^* \in C^{1,\alpha}(\Omega_{u^*}; \mathbb{R}^N) \quad \text{for any } \alpha \in (0,1) .$$

Of course Theorem 2.17 implies

Corollary 2.18. *With the notation and with the assumptions of Theorem 2.17 let us suppose that we have in addition to Assumption 2.1*

$$0 < D^2 f(Z)(Y,Y) \quad \text{for all } Z, Y \in \mathbb{R}^{nN} , \quad Y \neq 0 .$$

Then there exists an open set Ω_0 *of full measure, i.e.* $|\Omega - \Omega_0| = 0$, *such that*

$$u^* \in C^{1,\alpha}(\Omega_0; \mathbb{R}^N) \quad \text{for any } \alpha \in (0,1) .$$

Before proving Theorem 2.17 we introduce (following for example [GS], [AD] or [Bu]) a relaxed functional $\hat{J}$ on the space of functions of bounded variation and recall some well known properties (see Appendix A.1 for a more intensive discussion). The theorem then will be established as an immediate consequence of Lemma B.5 and Theorem 1.1 of [AG2] (compare Remark 2.16).

Here and below $\hat{\Omega}$ always denotes a bounded Lipschitz domain.

Definition 2.19. *For all* $w \in BV(\hat{\Omega}; \mathbb{R}^N)$ *the functional* $\hat{J}[w; \hat{\Omega}]$ *is given by*

$$\hat{J}[w; \hat{\Omega}] := \inf \left\{ \liminf_{k \to \infty} J[w_k] : w_k \in C^1(\hat{\Omega}; \mathbb{R}^N), \ w_k \to w \text{ in } L^1_{loc}(\hat{\Omega}; \mathbb{R}^N) \right\} .$$

The following properties of $\hat{J}$ are needed in our context:

Proposition 2.20.
 i) *The functional $\hat{J}$ is lower semicontinuous with respect to $L^1_{loc}(\hat{\Omega}; \mathbb{R}^N)$-convergence.*
 ii) *The functionals $\hat{J}[\cdot, \hat{\Omega}]$ and $J_{|\hat{\Omega}}$ coincide on $W^1_1(\hat{\Omega}; \mathbb{R}^N)$.*
 iii) *With the notation of Theorem 2.17, we have*

$$\hat{J}[u^*; \Omega] \leq \inf \left\{ J[u] : u \in u_0 + \overset{\circ}{W}^1_1(\Omega; \mathbb{R}^N) \right\}.$$

Proof. Using L^1-approximations for BV-functions, the first part immediately follows from the definition of $\hat{J}$. The lower semicontinuity of J on $W^1_1(\hat{\Omega}; \mathbb{R}^N)$ with respect to the L^1_{loc}-topology (see [AD]) implies part (ii.). To prove the third statement, consider a sequence $\{u_m\}$ and u^* as in Theorem 2.17. Here we assume without loss of generality that $u_m \in C^1(\Omega; \mathbb{R}^N)$ for all $m \in \mathbb{N}$. The strong L^1-convergence $u_m \to u$ yields

$$\hat{J}[u^*; \Omega] = \inf \left\{ \liminf_{k \to \infty} J[w_k] : w_k \in C^1(\Omega; \mathbb{R}^N), \right.$$

$$\left. w_k \to u^* \text{ in } L^1_{loc}(\Omega; \mathbb{R}^N) \right\}$$

$$\leq \liminf_{m \to \infty} J[u_m] = \inf \left\{ J[u] : u \in u_0 + \overset{\circ}{W}^1_1(\Omega; \mathbb{R}^N) \right\},$$

and the proposition is proved. $\square$

A deeper result is the following representation formula of Goffman and Serrin (see [GS]):

Proposition 2.21. *The representation formula*

$$\hat{J}[u, \hat{\Omega}] = \int_{\hat{\Omega}} f(\nabla^a u) \, dx + \int_{\hat{\Omega}} f_\infty \left(\frac{\nabla^s u}{|\nabla^s u|} \right) d|\nabla^s u|$$

holds for all $u \in BV(\hat{\Omega}; \mathbb{R}^N)$ where f_∞ is the recession function of f defined by

$$f_\infty(X) = \limsup_{t \to +\infty} \frac{f(tX)}{t}.$$

As above, the absolutely continuous part of ∇u with respect to the Lebesgue measure is denoted by $\nabla^a u$, the singular part by $\nabla^s u$ and $\nabla^s u / |\nabla^s u|$ is the Radon-Nikodym derivative.

For a proof we also refer to [AD], where f is only required to be quasiconvex.

The next proposition follows from [AG2], Theorem 2.1 and Proposition 2.2, respectively (see also [GMS1] and [Re]).

Proposition 2.22. *Suppose that there is $u \in BV(\hat{\Omega}; \mathbb{R}^N)$ and that there is a sequence $\{u_m\} \subset W^1_1(\hat{\Omega}; \mathbb{R}^N)$ such that as $m \to \infty$:*

i) $u_m \to u$ in $L^1(\hat{\Omega}; \mathbb{R}^N)$;

ii) $\int_{\hat{\Omega}} \sqrt{1 + |\nabla u_m|^2} \, dx \to \int_{\hat{\Omega}} \sqrt{1 + |\nabla u|^2}$.

Then $\hat{J}$ is continuous with respect to this kind of convergence, i.e.

$$\hat{J}[u_m; \hat{\Omega}] \to \hat{J}[u; \hat{\Omega}] \quad as \ m \to \infty.$$

We now come to the *Proof of Theorem 2.17.* Consider a sequence $\{u_m\} \subset u_0 + \overset{\circ}{W}{}^1_1(\Omega; \mathbb{R}^N)$ such that

$$J[u_m] \to \inf \left\{ J[u] : u \in u_0 + \overset{\circ}{W}{}^1_1(\Omega; \mathbb{R}^N) \right\}$$

and $u_m \to u^*$ in $L^1(\Omega; \mathbb{R}^N)$ as $m \to \infty$.

We improve the properties of $\{u_m\}$ by using Lemma B.5, i.e. given $x_0 \in \Omega$ we choose a ball $B_R(x_0)$, $B_{2R}(x_0) \Subset \Omega$ and we choose a sequence $\{w_m\} \subset u_0 + \overset{\circ}{W}{}^1_1(\Omega; \mathbb{R}^N)$ such that:

i) $w_m \to u^*$ in $L^1(\Omega; \mathbb{R}^N)$ as $m \to \infty$;

ii) $\lim\limits_{m \to \infty} J[w_m] \le \lim\limits_{m \to \infty} J[u_m]$;

iii) $w_{m|\partial B_R(x_0)} = u^*_{|\partial B_R(x_0)}$, where the traces are well defined functions of class $L^1(\partial B_R(x_0); \mathbb{R}^N)$.

We then let

- $I : W^1_1(B_R(x_0); \mathbb{R}^N) \to \mathbb{R}, \quad I[w] := \int_{B_R(x_0)} f(\nabla w) \, dx,$

- $\mathbb{K} := \left\{ w \in W^1_1(B_R(x_0); \mathbb{R}^N) : w_{|\partial B_R(x_0)} = u^*_{|\partial B_R(x_0)} \right\},$

and claim that (using the notation $v_m = w_{m|B_R(x_0)}$)

$$\inf_{\mathbb{K}} I = \liminf_{m \to \infty} I[v_m]. \tag{24}$$

In fact, we argue by contradiction and assume that there is $w \in \mathbb{K}$ satisfying

$$I[w] \le I[v_m] - \delta$$

for some $\delta > 0$ and for all m sufficiently large. By the definition of $\mathbb{K}$, the function

$$\tilde{w}_m := \begin{cases} w & \text{on } B_R(x_0) \\ w_m & \text{on the rest of } \Omega \end{cases}$$

is of class $W^1_1(\Omega; \mathbb{R}^N)$ and, as a direct consequence, we have

$$\limsup_{m\to\infty} J[\tilde{w}_m] \leq \lim_{m\to\infty} J[w_m] - \delta \,.$$

This provides a contradiction to $ii)$, hence we have (24).

Now we complete the proof of Theorem 2.17: to this purpose consider $\tilde{u} \in BV(B_R(x_0); \mathbb{R}^N)$ such that $\mathrm{spt}(\tilde{u} - u^*) \Subset B_R(x_0)$. It is well known (and for the reader's convenience we give a proof in Lemma B.1) that we may choose $\tilde{u}_m \in W_1^1(B_R(x_0); \mathbb{R}^N)$, $m \in \mathbb{N}$, satisfying $\tilde{u}_m \to \tilde{u}$ in $L^1(B_R(x_0); \mathbb{R}^N)$,

$$\int_{B_R(x_0)} \sqrt{1 + |\nabla \tilde{u}_m|^2} \, dx \to \int_{B_R(x_0)} \sqrt{1 + |\nabla \tilde{u}|^2}$$

and $\tilde{u}_{m|\partial B_R(x_0)} = \tilde{u}_{|\partial B_R(x_0)}$. Then we have $\tilde{u}_m \in \mathbb{K}$, hence, by Proposition 2.20, by Proposition 2.22 and by (24)

$$\hat{J}[\tilde{u}; B_R(x_0)] = \lim_{m\to\infty} \hat{J}[\tilde{u}_m; B_R(x_0)] = \lim_{m\to\infty} I[\tilde{u}_m]$$

$$\geq \inf_{\mathbb{K}} I = \liminf_{m\to\infty} I[v_m] \geq \hat{J}[u^*; B_R(x_0)] \,, \tag{25}$$

where the last inequality follows from $v_m \to u^*$ in $L^1(\Omega; \mathbb{R}^N)$ (see $i)$) and from the definition of $\hat{J}[\cdot; B_R(x_0)]$. Quoting Theorem 1.1 of [AG2] we have proved Theorem 2.17. $\square$

Remark 2.23. *In particular, the conclusion of Theorem 2.17 holds for each L^1-limit of the δ-regularization given in Section 2.1.2.*

2.3.2 Partial $C^{0,\alpha}$-regularity of the dual solution

Now the second partial regularity result is proved, i.e. we concentrate on the dual solution σ. We prefer to give a "direct" proof just relying on our δ-regularization, an alternative way is outlined in [SE1/4/5/6]: by using the relaxed minimax inequality one can show that $\nabla u^* = \nabla f^*(\sigma)$ holds on the regular set of any cluster point u^* of a J-minimizing sequence. As a consequence, we have $\sigma = \nabla f(\nabla u^*)$ which implies Theorem 2.24.

Theorem 2.24. *Suppose that we have Assumption 2.1 and let u^* denote a weak cluster point of the δ-regularization introduced in Section 2.1.2. Moreover, consider the solution σ of the dual variational problem $(\mathcal{P}^*)$. Then*

$$\sigma \in C^{0,\alpha}(\Omega_{u^*}; \mathbb{R}^{nN}) \quad \text{for any } 0 < \alpha < 1 \,,$$

where Ω_{u^*} is the open set given above.

Again we immediately obtain

Corollary 2.25. *If in addition to the assumptions of Theorem 2.24*

$$0 < D^2 f(Z)(Y, Y) \quad \text{for all } Z, Y \in \mathbb{R}^{nN} \,, \; Y \neq 0$$

is assumed, then partial $C^{0,\alpha}$-regularity of σ follows on an open set of full measure.

Proof of Theorem 2.24. Consider the δ-regularization introduced in Section 2.1.2 and fix a subsequence with L^1-cluster point u^*. Recalling $\sigma_\delta := \delta \nabla u_\delta + \nabla f(\nabla u_\delta)$ we have the equation

$$\int_\Omega \sigma_\delta : \nabla \varphi \, dx = 0 \quad \text{for all } \varphi \in C_0^1(\Omega; \mathbb{R}^N) \,. \tag{26}$$

By Theorem 2.17 we know that u^* satisfies $u^* \in C^{1,\alpha}(\Omega_{u^*}; \mathbb{R}^N)$ and that Ω_{u^*} is an open set. Now, for every open set G such that $\overline{G} \subset \Omega_{u^*}$ we have $\nabla u^* \in L^\infty(G; \mathbb{R}^{nN})$ and additionally

$$\int_G \nabla f(\nabla u^*) : \nabla \varphi \, dx = 0 \quad \text{for all } \varphi \in C_0^1(G; \mathbb{R}^N) \,. \tag{27}$$

This is the Euler equation with respect to the relaxation $\hat{J}$: by (25) u^* is a local minimizer of $\hat{J}[\cdot, B_R(x)]$ for a suitable ball around each point $x \in \Omega$. The representation formula from Proposition 2.21 then shows that variations on the regular set imply (27).

Combining the Euler equations (26) and (27) we obtain (passing to a subsequence)

$$\nabla u_\delta(x) \to \nabla u^*(x) \quad \text{for almost every } x \in G \text{ as } \delta \to 0. \tag{28}$$

In fact, inserting $\eta^2[u_\delta - u^*]$, $\eta \in C_0^1(G)$, $0 \le \eta \le 1$, and subtracting the equations we get

$$\int_G (\sigma_\delta - \nabla f(\nabla u^*)) : \nabla(\eta^2[u_\delta - u^*]) \, dx = 0 \,.$$

The definition of σ_δ yields

$$\begin{aligned}
&\int_G (\nabla f(\nabla u_\delta) - \nabla f(\nabla u^*)) : (\nabla u_\delta - \nabla u^*)\eta^2 \, dx \\
&+ \int_G \delta \nabla u_\delta : (\nabla u_\delta - \nabla u^*)\eta^2 \, dx \\
&= -2 \int_G \sigma_\delta : ([u_\delta - u^*] \otimes \nabla \eta)\, \eta \, dx \\
&+ 2 \int_G \nabla f(\nabla u^*) : ([u_\delta - u^*] \otimes \nabla \eta)\, \eta \, dx \,.
\end{aligned} \tag{29}$$

Now, $\nabla f(\nabla u^*) \in L^\infty(\Omega; \mathbb{R}^{nN})$ and $(u_\delta - u^*) \to 0$ in $L^1(\Omega; \mathbb{R}^{nN})$, hence

$$\int_G \nabla f(\nabla u^*) : ([u_\delta - u^*] \otimes \nabla \eta)\, \eta \, dx \to 0 \quad \text{as } \delta \to 0 \,.$$

Moreover,

$$\int_G \delta \nabla u_\delta : (\nabla u_\delta - \nabla u^*)\eta^2 \, dx \to 0 \quad \text{as } \delta \to 0$$

follows from $\delta \nabla u_\delta \to 0$ in $L^2(\Omega; \mathbb{R}^{nN})$ and $\delta \int_\Omega |\nabla u_\delta|^2 \, dx \to 0$ (compare (9), Section 2.1.2). The first integral on the right-hand side of (29) can be written in the following form

$$-2 \int_G \delta \nabla u_\delta : ([u_\delta - u^*] \otimes \nabla \eta)\, \eta \, dx$$

$$-2 \int_G \nabla f(\nabla u_\delta) : ([u_\delta - u^*] \otimes \nabla \eta)\, \eta \, dx =: I_1 + I_2 \, .$$

As above we immediately get $I_2 \to 0$ as $\delta \to 0$ and the same is true for I_1. This implies

$$\int_G (\nabla f(\nabla u_\delta) - \nabla f(\nabla u^*)) : (\nabla u_\delta - \nabla u^*)\eta^2 \, dx \to 0 \quad \text{as } \delta \to 0 \, .$$

Passing to a subsequence we get

$$(\nabla f(\nabla u_\delta) - \nabla f(\nabla u^*)) : (\nabla u_\delta - \nabla u^*) \to 0 \text{ almost everywhere in } G \quad (30)$$

as $\delta \to 0$. For the sake of completeness we next briefly prove a remark which is needed to deduce (28) from (30).

Remark 2.26. *For any $Z \in \mathbb{R}^{nN}$ we have*

$$\lim_{|Y| \to \infty} (\nabla f(Y) - \nabla f(Z)) : (Y - Z) = \infty \, .$$

Proof. Fix $Z \in \mathbb{R}^{nN}$. For any $Y \in \mathbb{R}^{nN}$ the convexity of f implies

$$(\nabla f(Y) - \nabla f(Z)) : (Y - Z) \geq f(Y) - f(Z) - \nabla f(Z) : (Y - Z)$$

$$= f(Y) - \nabla f(Z) : Y + c(Z) \, ,$$

where $c(Z)$ is a real number depending on Z. The definition of the conjugate function gives the generalized Young's inequality

$$\nabla f(Z) : Y \leq f((1 - \varepsilon)Y) + f^*\left((1 - \varepsilon)^{-1}\nabla f(Z)\right)$$

for any real number $0 < \varepsilon < 1$. By the open mapping theorem $\text{Im}(\nabla f)$ is an open set and we may choose $\varepsilon > 0$ sufficiently small such that in addition $f^*\left((1 - \varepsilon)^{-1}\nabla f(Z)\right) < c(Z)$ can be assumed. Setting $Q := (1 - \varepsilon)Y$ it remains to estimate

$$f(Y) - f((1 - \varepsilon)Y) \geq \varepsilon Y : \nabla f((1 - \varepsilon)Y) = \frac{\varepsilon}{1 - \varepsilon} Q : \nabla f(Q)$$

$$\geq \frac{\varepsilon}{1 - \varepsilon} (f(Q) - f(0))$$

which follows from the convexity of f. □

From Remark 2.26 and from (30) we deduce that $\nabla u_\delta(x)$ remains bounded for almost every $x \in G$, but then (30) immediately yields (28) by the strict convexity of f and we have

$$\nabla f\big(\nabla u_\delta(x)\big) \to \nabla f\big(\nabla u^*(x)\big) \quad \text{for almost every } x \in G \text{ as } \delta \to 0.$$

Keeping (28) and $\nabla f(\nabla u_\delta) \to \sigma$ in $L^2(\Omega; \mathbb{R}^{nN})$ as $\delta \to 0$ in mind, it is shown that

$$\sigma(x) = \nabla f\big(\nabla u^*(x)\big) \quad \text{in } G. \tag{31}$$

This implies (31) for $x \in \Omega_{u^*}$ and the theorem is proved. □

2.4 Degenerate variational problems with linear growth

Let us recall the main line of the last section where we proved a smoothness result for the dual solution: the first step is to show that generalized minimizers $u^* \in \mathcal{M}$ are regular on Ω_{u^*} (see Section 2.3.1). The essential tool to analyze the maximizer σ with the help of this information turns out to be the duality relation proved in (31) of Section 2.3.2

$$\sigma(x) = \nabla f\big(\nabla u^*(x)\big) \quad \text{for all } x \in \Omega_{u^*}. \tag{32}$$

If degenerate problems are studied, i.e. if $|\Omega - \Omega_{u^*}| > 0$ has to be expected, then the approach outlined above in general does not lead to satisfying results. Let us sketch the two main problems by considering a prominent example (compare [GMS1]):

$$f(Z) = \big(1 + |Z|^k\big)^{\frac{1}{k}}, \quad k > 2. \tag{33}$$

On one hand, (32) now is a quite vague statement since, due to the degeneracy of $D^2 f$, the regular set Ω_{u^*} may be very small.

On the other hand, and this is even more restrictive if we are interested in regularity results for σ, partial Hölder continuity of σ for the integrand (33) at hand follows a priori only on the set $\big[D^2 f(\nabla u^*(x)) > 0\big] = \big[\nabla u^*(x) \neq 0\big]$. However, an intrinsic theorem should be formulated in terms of σ, i.e. the domain of partial regularity is expected to be

$$\big[\sigma(x) \neq 0\big] \supset \big[\nabla^a u^*(x) \neq 0\big]. \tag{34}$$

Here the inclusion follows on account of (32) and of course is meant modulo sets of measure zero. Again $\nabla^a u^*$ denotes the absolutely continuous part of ∇u^* with respect to the Lebesgue-measure and $\nabla^s u^*$ will be used as the symbol for the singular part.

In this section, a generalization of the classical duality relation is established for almost all $x \in \Omega$: we leave the regularity of u^* as the starting point and prove by arguments from measure theory (in addition we use Appendix B.3) that there is a generalized minimizer $u^* \in \mathcal{M}$ of the problem $(\mathcal{P})$ such that

$$\sigma(x) = \nabla f(\nabla^a u^*(x)) \quad \text{for almost all } x \in \Omega .$$

Here, the degenerate situation $D^2 f \geq 0$ also is covered (compare Assumption 2.27). Coming back to the above example, we see that in fact equality (again modulo sets of measure zero) holds in (34) and analogous results of course are true in the case of more general degenerate integrands. As an application, an intrinsic regularity theorem independent of u^* can be formulated just in terms of σ and the data (compare Section 2.4.2 for details).

2.4.1 The duality relation for degenerate problems

In order to establish the duality relation (32) in the degenerate situation, the general Assumption 2.1 is refined by the additional

Assumption 2.27. *Assume in addition to Assumption 2.1 that*

i) $f(Z) > 0$ *for all* $Z \in \mathbb{R}^{nN}$, $Z \neq 0$, *and* $f(0) = 0$.

ii) We have for all $Z, Y \in \mathbb{R}^{nN}$, $Z \neq 0$, $Y \neq 0$,

$$0 < D^2 f(Z)(Y, Y) .$$

Remark 2.28.
i) Of course i) is supposed without loss of generality and the model integrand considered in (33) satisfies ii).
ii) The set Ω_{u^*} is characterized by

$$\lim_{r \to 0} \frac{1}{|B_r(x)|} \int_{B_r(x)} |\nabla^a u^* - P| \, dx + \lim_{r \to 0} \frac{1}{|B_r(x)|} \int_{B_r(x)} |\nabla^s u^*| = 0$$

for some matrix $P \in \mathbb{R}^{nN}$ such that $D^2 f(P) > 0$ (see [AG2] and [BF1]). Hence, we have $\nabla^a u^* = 0$ almost everywhere on the complement $\Omega_{u^*}^c$ of Ω_{u^*}. However, this provides no results at all because no topological information on $\Omega_{u^*}^c$ is available. Moreover, the singular part $\nabla^s u^*$ is not necessarily vanishing on $\Omega_{u^*}^c$.

Let us recall the notion of the δ-regularization $\{u_\delta\}$ introduced in Section 2.1.2. In particular, the convergence (7), Section 2.1.2, and Theorem 2.10 are needed to assume without loss of generality (after passing to a subsequence)

$$i) \qquad \sigma_\delta(x) \to \sigma(x) \quad \text{for almost all } x \in \Omega ,$$

$$ii) \quad \delta \nabla u_\delta(x) \to 0 \qquad \text{for almost all } x \in \Omega , \tag{35}$$

where $\sigma_\delta = \delta \nabla u_\delta + \nabla f(\nabla u_\delta)$ and where σ denotes the unique solution of the dual variational problem $(\mathcal{P}^*)$. Passing to another subsequence, if necessary, a L^1-cluster point u^* of u_δ is fixed in the following:

$$u_\delta \xrightarrow{L^1} u^* \in BV(\Omega, \mathbb{R}^N) \quad \text{as } \delta \to 0 .$$

Then the main theorem of this section reads as follows.

Theorem 2.29. *The unique solution σ of the dual problem $(\mathcal{P}^*)$ satisfies*

$$\sigma(x) = \nabla f(\nabla^a u^*(x)) \quad \text{for almost all } x \in \Omega .$$

Remark 2.30. *It remains an open question whether $\sigma = \nabla f(\nabla^a u)$ holds for any generalized minimizer $u \in \mathcal{M}$.*

For the proof of Theorem 2.29 we have to construct "large" sets of uniform convergence according to the following Proposition.

Proposition 2.31. *There is a measurable function $v\colon \Omega \to \mathbb{R}^{nN}$, and for any $\varepsilon > 0$ there is a compact set $K \Subset \Omega$ such that:*

i) $\sigma_\delta \rightrightarrows \sigma$ on K, $\sigma(x) \notin \partial\mathrm{Im}(\nabla f)$ for all $x \in K$;

ii) $\delta \nabla u_\delta \rightrightarrows 0$ on K;

iii) $\nabla u_\delta \rightrightarrows v$ on K;

iv) The restriction of v on K is a continuous function;

v) $|\Omega - K| < \varepsilon$.

Remark 2.32. *In the following it is obvious that we can restrict ourselves to the consideration of Lebesgue points of σ and ∇u^*, respectively. This is always assumed as a general "hypothesis".*

Proof of Proposition 2.31. Let us first define

$$v_j^i(x) := \limsup_{\delta \to 0} \frac{\partial}{\partial x_j} u_\delta^i(x) , \quad i \in \{1,\ldots,N\} , \quad j \in \{1,\ldots,n\} ,$$

which by construction is a measurable function with values in $\overline{\mathbb{R}}$.

Now fix $\varepsilon > 0$. The uniform convergence (as stated in i) and ii)) on a compact set $\tilde{K} \Subset \Omega$ with $|\Omega - \tilde{K}| < \varepsilon/2$ follows on account of (35) and Egoroff's theorem. Setting $N = \{x \in \Omega : \sigma(x) \in \partial\mathrm{Im}(\nabla f)\}$ it was proved in Section 2.2 that $|N| = 0$. Hence, we may choose an open set $U \supset N$ with $|U| < \varepsilon/2$. Then $K := \tilde{K} - U$ is a compact set such that i) and ii) are valid and such that in addition K satisfies v).

Next observe that we have on K as $\delta \to 0$

$$\nabla f(\nabla u_\delta(x)) = \sigma_\delta(x) - \delta \nabla u_\delta(x) \to \sigma(x) . \tag{36}$$

If $x_0 \in K$ is fixed, then $\sigma(x_0) \notin \partial \mathrm{Im}(\nabla f)$ implies that there is a constant $\rho = \rho(x_0)$ such that for all δ sufficiently small

$$\mathrm{dist}\Big(\nabla f(\nabla u_\delta(x_0)), \partial \mathrm{Im}(\nabla f)\Big) \geq \rho .$$

This means that we have for all δ sufficiently small

$$\nabla f(\nabla u_\delta(x_0)) \in C := \Big\{Q \in \mathrm{Im}(\nabla f) : \mathrm{dist}(Q, \partial \mathrm{Im}(\nabla f)) \geq \rho\Big\} .$$

Since C is compact and since ∇f is a homeomorphism (compare the proof of Lemma 2.14), $(\nabla f)^{-1}(C)$ is compact, in particular $(\nabla f)^{-1}(C)$ is bounded and, as a consequence,

$$\limsup_{\delta \to 0} |\nabla u_\delta(x)| < \infty \quad \text{for all } x \in K . \tag{37}$$

Given (36) and (37), the pointwise convergence of $\nabla u_\delta(x)$ on K is clear since ∇f is one-to-one. Egoroff's Theorem then proves $iii)$ on a "large" compact set $\tilde{K} \subset K$, without loss of generality on K. The proof of $iv)$ is an application of Lusin's theorem and the proposition follows. $\square$

We now come to the *Proof of Theorem 2.29*. Fix $\varepsilon > 0$ and choose K according to the Proposition 2.31. Since K is a measurable set, the Lebesgue-Besicovitch Differentiation Theorem yields

$$\lim_{r \to 0} \frac{|B_r(x) \cap K|}{|B_r(x)|} = 1 \quad \text{for almost all } x \in K . \tag{38}$$

It is also known that for almost all $x \in \Omega$ there exists a matrix $P \in \mathbb{R}^{nN}$ such that

$$\lim_{r \to 0} \frac{1}{|B_r(x)|} \int_{B_r(x)} |\nabla u^* - P|$$
$$:= \lim_{r \to 0} \frac{1}{|B_r(x)|} \int_{B_r(x)} |\nabla^a u^* - P| \, dx + \lim_{r \to 0} \frac{1}{|B_r(x)|} \int_{B_r(x)} |\nabla^s u^*| = 0 .$$

Let us first consider the case $P \neq 0$, i.e. x is a non-degenerate point. Going through the lines of Section 2.3 we observe that the duality relation as claimed in the theorem holds at this particular point x. Thus, we have to study the case

$$\lim_{r \to 0} \frac{1}{|B_r(x)|} \int_{B_r(x)} |\nabla u^*| = 0 . \tag{39}$$

Observe that (39) implies

$$\lim_{r \to 0} \frac{1}{|B_r(x)|} \int_{B_r(x)} f(\nabla^a u^*) \, dx = 0 . \tag{40}$$

In fact, on account of the continuity of f and since $f(0) = 0$ we may fix a real number $\lambda > 0$ and find $\kappa > 0$ such that $|Z| < \kappa$ implies $f(Z) < \lambda$. We obtain

$$
\lim_{r \to 0} \frac{1}{|B_r(x)|} \int_{B_r(x)} f(\nabla^a u^*) \, dx
$$

$$
\leq \limsup_{r \to 0} \frac{1}{|B_r(x)|} \int_{B_r(x) \cap [|\nabla^a u^*| < \kappa]} f(\nabla^a u^*) \, dx
$$

$$
+ \limsup_{r \to 0} \frac{1}{|B_r(x)|} \int_{B_r(x) \cap [|\nabla^a u^*| > \kappa]} f(\nabla^a u^*) \, dx \, .
$$

Here, the second term on the right-hand side vanishes by the linear growth of f and by (39): for some real numbers c_1, $c_2 > 0$ we have

$$
\limsup_{r \to 0} \frac{1}{|B_r(x)|} \int_{B_r(x) \cap [|\nabla^a u^*| > \kappa]} f(\nabla^a u^*) \, dx
$$

$$
\leq c_1 \limsup_{r \to 0} \frac{1}{|B_r(x)|} \int_{B_r(x) \cap [|\nabla^a u^*| > \kappa]} |\nabla^a u^*| \, dx
$$

$$
+ c_2 \limsup_{r \to 0} \frac{1}{|B_r(x)|} \int_{B_r(x) \cap [|\nabla^a u^*| > \kappa]} 1 \, dx = 0 \, .
$$

The remaining term is bounded from above by λ which is an arbitrary fixed positive number, and the claim (40) is proved.

With (38) and (39) we define a set G_K satisfying $|K - G_K| = 0$:

$$
G_K := \left\{ x \in K : \text{(38) and (39) are valid or } \sigma(x) = \nabla f(\nabla^a u^*(x)) \right\} .
$$

Remark 2.33. *Let us give a short comment on this definition. Consider the set of all $x \in K$ such that (38) holds. Then we have to distinguish between the cases "$P = 0$", i.e. (39) is true, and "$P \neq 0$". As already mentioned, in the second case the duality relation is proved in Section 2.3.2, in particular $|K - G_K| = 0$.*

According to this remark, we fix $\hat{x} \in G_K$ satisfying (38) and (39) and recall the fact that f achieves its absolute minimum at $Z = 0$. Hence, $0 = \nabla^a u^*(\hat{x})$ and $0 = \nabla f(0)$. We claim that

$$
\sigma(\hat{x}) = 0 = \nabla f(0) = \nabla f(\nabla^a u^*(\hat{x})) \tag{41}
$$

which immediately yields the theorem by passing to the limit $\varepsilon \to 0$.

To prove (41) assume by contradiction that $\sigma(\hat{x}) \neq 0$. We now claim that there is a real number $\gamma = \gamma(\hat{x}) > 0$ such that for all δ sufficiently small

$$
\gamma < |\nabla u_\delta(\hat{x})| \, . \tag{42}
$$

To verify (42), let $\tau = |\sigma(\hat{x})|$ and choose $\delta_0 > 0$ sufficiently small. We obtain for all $\delta < \delta_0$

$$|\sigma_\delta(\hat{x}) - \sigma(\hat{x})| < \frac{\tau}{2} \quad \text{and} \quad |\delta \nabla u_\delta(\hat{x})| < \frac{\tau}{4} .$$

This gives for all $\delta < \delta_0$

$$|\nabla f(\nabla u_\delta(\hat{x}))| \geq |\sigma_\delta(\hat{x})| - |\delta \nabla u_\delta(\hat{x})| > \frac{\tau}{4} . \tag{43}$$

If it is supposed in contradiction to (42) that there is a sequence $\{\delta_n\}$, $\delta_n \to 0$ as $n \to \infty$, such that

$$\nabla u_{\delta_n}(\hat{x}) \to 0 \quad \text{as } n \to \infty ,$$

then the continuity of ∇f yields

$$\nabla f(\nabla u_{\delta_n}(\hat{x})) \to \nabla f(0) = 0 \quad \text{as } n \to \infty .$$

This, however, is excluded by (43) and (42) is proved.

By Proposition 2.31, $iii)$, it also follows that $\gamma \leq |v(\hat{x})|$. Thus, by continuity of v on K, there is a real number $\rho_0 > 0$ such that $B_{4\rho_0}(\hat{x}) \Subset \Omega$ and such that for any $\rho < \rho_0$

$$\frac{\gamma}{2} \leq |v(x)| \quad \text{for all } x \in B_\rho(\hat{x}) \cap K .$$

Finally, setting $\kappa = \gamma/4$ and recalling the uniform convergence stated in Proposition 2.31, $iii)$, we decrease δ_0, if necessary, and arrive at

$$\kappa \leq |\nabla u_\delta(x)| \quad \text{for all } x \in B_\rho(\hat{x}) \cap K , \quad 0 < \rho < \rho_0 , \tag{44}$$

for all $\delta < \delta_0$.

Remark 2.34. *If in the sense of measures*

$$|\nabla u_\delta| \rightharpoonup |\nabla u^*| \tag{45}$$

would be known, then the compactness of $\overline{B}_\rho(\hat{x}) \cap K$ would imply

$$\kappa |\overline{B}_\rho(\hat{x}) \cap K| \leq \limsup_{\delta \to 0} |\nabla u_\delta|(\overline{B}_\rho(\hat{x}) \cap K)$$

$$\leq |\nabla u^*|(\overline{B}_\rho(\hat{x}) \cap K) .$$

Passing to the limit $\rho \to 0$ a contradiction would follow from (39) and from the density relation (38).

Hence, we have to establish an appropriate substitute for (45) where we benefit from the "minimality" of $\{u_\delta\}$. As in Section 2.3.1 (see in particular (24) and (25)) we modify $\{u_\delta\}$ following Lemma B.5, i.e.: we choose for almost any ρ as above a sequence $\{w_m\} \subset u_0 + \overset{\circ}{W}{}^1_1(\Omega; \mathbb{R}^N)$, $\tilde{w}_m := w_{m|B_{2\rho}(\hat{x})}$, satisfying *i)* $w_m \to u^*$ in $L^1(\Omega; \mathbb{R}^N)$ as $m \to \infty$;

ii) $w_{m|\partial B_{2\rho}(\hat{x})} = u^*_{|\partial B_{2\rho}(\hat{x})}$;

iii) $\displaystyle \liminf_{m \to \infty} I[\tilde{w}_m] = \inf_{\mathbb{K}} I = \hat{J}[u^*; B_{2\rho}(\hat{x})]$;

iv) $\tilde{w}_{m|B_\rho(\hat{x})} = u_{\delta_m|B_\rho(\hat{x})}$.

Here $\{u_{\delta_m}\}$ denotes a subsequence of $\{u_\delta\}$ and analogous to Section 2.3.1 we set

- $I : W_1^1\big(B_{2\rho}(\hat{x}); \mathbb{R}^N\big) \to \mathbb{R}$, $I[w] := \displaystyle\int_{B_{2\rho}(\hat{x})} f(\nabla w)\, \mathrm{d}x$,

- $\mathbb{K} := \Big\{ w \in W_1^1\big(B_{2\rho}(\hat{x}); \mathbb{R}^N\big) : w_{|\partial B_{2\rho}(\hat{x})} = u^*_{|\partial B_{2\rho}(\hat{x})} \Big\}$.

Now, the convexity of f and Assumption 2.27, *i)*, imply the existence of a real number $\vartheta > 0$ such that $f(Z) > \vartheta$ if $|Z| \geq \kappa$. Hence, we deduce from (44) and *iv)*

$$\vartheta \leq f(\nabla \tilde{w}_m) \quad \text{for all } x \in B_\rho(\hat{x}) \cap K$$

and for all $m \in \mathbb{N}$. This yields (recall $f \geq 0$, see *iii)* and Proposition 2.21)

$$\begin{aligned}
\vartheta \frac{|K \cap B_\rho(\hat{x})|}{|B_{2\rho}(\hat{x})|} &\leq \frac{1}{|B_{2\rho}(\hat{x})|} \liminf_{m \to \infty} \int_{B_{2\rho}(\hat{x})} f(\nabla \tilde{w}_m)\, \mathrm{d}x \\
&= \frac{1}{|B_{2\rho}(\hat{x})|} \inf_{\mathbb{K}} I = \frac{1}{|B_{2\rho}(\hat{x})|} \hat{J}\big[u^*; B_{2\rho}(\hat{x})\big] \\
&= \frac{1}{|B_{2\rho}(\hat{x})|} \int_{B_{2\rho}(\hat{x})} f(\nabla^a u^*)\, \mathrm{d}x \\
&\quad + \frac{1}{|B_{2\rho}(\hat{x})|} \int_{B_{2\rho}(\hat{x})} f_\infty\left(\frac{\nabla^s u^*}{|\nabla^s u^*|}\right) d|\nabla^s u^*|\,.
\end{aligned} \tag{46}$$

Both sides of (46) are independent of m and we now may pass to the limit $\rho \to 0$. The density assumption (38) implies

$$\lim_{\rho \to 0} \vartheta \frac{|K \cap B_\rho(\hat{x})|}{|B_{2\rho}(\hat{x})|} = \vartheta 2^{-n}\,, \tag{47}$$

whereas on account of (39), (40) and the boundedness of f_∞

$$\lim_{\rho \to 0} \left\{ \frac{1}{|B_{2\rho}(\hat{x})|} \int_{B_{2\rho}(\hat{x})} f(\nabla^a u^*)\, \mathrm{d}x \right. \\
\left. + \frac{1}{|B_{2\rho}(\hat{x})|} \int_{B_{2\rho}(\hat{x})} f_\infty\left(\frac{\nabla^s u^*}{|\nabla^s u^*|}\right) d|\nabla^s u^*| \right\} = 0\,. \tag{48}$$

Thus, (47) and (48) contradict (46) and Theorem 2.29 is proved. $\square$

2.4.2 Application: an intrinsic regularity theory for σ

We finish this section with a short application of Theorem 2.29 to the regularity theory: the u^*- and the σ-degenerate sets are identified modulo sets of measure zero and, as a consequence, an intrinsic regularity theorem for σ is obtained for the degenerate problems under consideration. To this purpose consider u^* as given above and let

$$\Omega_{u^*}^{\mathrm{deg}} := \left\{ x \in \Omega : \nabla^a u^*(x) = 0 \right\},$$

$$\Omega_{\sigma}^{\mathrm{deg}} := \left\{ x \in \Omega : \sigma(x) = \nabla f(0) = 0 \right\}.$$

The sets $\Omega_{u^*}^{\mathrm{deg}}$ and $\Omega_{\sigma}^{\mathrm{deg}}$ are well defined on the complements of sets of measure zero. For a more precise definition one has to consider Lebesgue points of σ and $\nabla^a u^*$, respectively, where the singular part $\nabla^s u^*$ should vanish. Since ∇f is one-to-one Theorem 2.29 implies:

Corollary 2.35. *With the assumptions of this section there exists a generalized minimizer $u^* \in \mathcal{M}$ such that*

$$\left| \Omega_{u^*}^{\mathrm{deg}} - \Omega_{\sigma}^{\mathrm{deg}} \right| = \left| \Omega_{\sigma}^{\mathrm{deg}} - \Omega_{u^*}^{\mathrm{deg}} \right| = 0.$$

On the other hand, the results of Section 2.3.1 imply: there is an open set $\Omega^{\mathrm{reg}} \subset \Omega - \Omega_{u^*}^{\mathrm{deg}}$ such that for any $0 < \alpha < 1$

$$u^* \in C^{1,\alpha}\left(\Omega^{\mathrm{reg}}; \mathbb{R}^N\right) \quad \text{and} \quad \left|(\Omega - \Omega_{u^*}^{\mathrm{deg}}) - \Omega^{\mathrm{reg}}\right| = 0.$$

By Theorem 2.29 (in fact (32) is sufficient), the dual solution σ is known to be of class $C^{0,\alpha}$ on Ω^{reg}. Now observe that, again by (32) and since ∇f is one-to-one, the inclusion $\Omega^{\mathrm{reg}} \subset \Omega - \Omega_{\sigma}^{\mathrm{deg}}$ is also valid. Applying Corollary 2.35 we get the following partial regularity result for σ.

Corollary 2.36. *If f is given as above and if σ denotes the unique solution of the dual variational problem $(\mathcal{P}^*)$, then there is an open set*

$$\Omega^{\mathrm{reg}} \subset \Omega - \Omega_{\sigma}^{\mathrm{deg}}$$

such that for any $0 < \alpha < 1$

$$\sigma \in C^{0,\alpha}\left(\Omega^{\mathrm{reg}}; \mathbb{R}^{nN}\right) \quad \text{and} \quad \left|(\Omega - \Omega_{\sigma}^{\mathrm{deg}}) - \Omega^{\mathrm{reg}}\right| = 0.$$

3

Variational integrands with (s, μ, q)-growth

In this chapter we extend the results mentioned in A and B of the table given in the introduction, i.e.: we concentrate on integrands with nearly linear and/or anisotropic growth together with some appropriate ellipticity condition.

In Section 3.1 we shortly discuss a basic existence result in Orlicz-Sobolev spaces.

Then we introduce the notion of (s, μ, q)-growth in Section 3.2 which is motivated by the following observation (compare also [Ma5]–[Ma7]). A theorem on the smoothness properties of solutions to our variational problem $(\mathcal{P})$ is expected to depend on three free parameters: the lower growth rate s of the integrand f which provides the starting integrability of the gradient of the solution. Moreover, we need some suitable assumptions on the exponents μ and q occurring in the non-uniform ellipticity condition.

We like to emphasize that this notion serves as an appropriate approach to the regularity theory of our variational problems at hand.

- The integrands discussed in the introduction are included within the notion of (s, μ, q)-growth.
- The class of admissible integrands is generalized to a large extent. In particular, the known results are recovered and substantially generalized.
- The variety of different settings discussed in the introduction is unified. Proving the results of this chapter, we do not have to consider several cases.
- The notion of (s, μ, q)-growth is adapted to the scalar situation as well as to the vector-valued setting.

Imposing the so-called (s, μ, q)-condition which relates the parameters in an appropriate way, the theorems of type (1) and (3) are proved in Section 3.3. Here we first show uniform local a priori gradient L^q-estimates for a suitable regularization. These a priori estimates are also valid in the general vectorial setting. The conclusion then follows from a DeGiorgi-type argument.

Partial regularity is obtained in Section 3.4 under the same hypotheses as given in Section 3.3. We benefit from the lemma on higher integrability mentioned above, and then argue using a blow-up procedure.

The comparison with the known results given in Section 3.5 completes the study of the general situation.

We finish this chapter by proving a theorem on the absence of singular points in the two-dimensional vector-valued case. Here, for the sake of simplicity, we restrict ourselves to energy densities with anisotropic (p, q)-growth, $1 < p \leq q$.

3.1 Existence in Orlicz-Sobolev spaces

Let us start with some basic definitions: suppose that we are given a N-function F having the Δ_2-property. This means (see, e.g. [Ad] for details) that the function $F \colon [0, \infty) \to [0, \infty)$ satisfies

$$F \text{ is continuous, strictly increasing and convex} ; \tag{N1}$$

$$\lim_{t \downarrow 0} \frac{F(t)}{t} = 0 , \quad \lim_{t \uparrow \infty} \frac{F(t)}{t} = +\infty ; \tag{N2}$$

$$\text{there exist } k, t_0 \geq 0 \text{ such that } F(2t) \leq k F(t) \text{ for all } t \geq t_0 . \tag{N3}$$

With (N1)–(N3) F generates the Orlicz space $L_F(\Omega; \mathbb{R}^N)$ equipped with the Luxemburg norm

$$\|u\|_{L_F(\Omega;\mathbb{R}^N)} := \inf \left\{ l > 0 : \int_\Omega F\left(\frac{1}{l}|u|\right) \, dx \leq 1 \right\} .$$

We then introduce the Orlicz-Sobolev space $W_F^1(\Omega; \mathbb{R}^N)$,

$$W_F^1(\Omega; \mathbb{R}^N) := \Big\{ u : \Omega \to \mathbb{R}^N \text{ such that } u \text{ is measurable and}$$

$$u \in L_F(\Omega; \mathbb{R}^N), \ \nabla u \in L_F(\Omega; \mathbb{R}^{nN}) \Big\} .$$

Together with the norm

$$\|u\|_{W_F^1(\Omega;\mathbb{R}^N)} = \|u\|_{L_F(\Omega;\mathbb{R}^N)} + \|\nabla u\|_{L_F(\Omega;\mathbb{R}^{nN})}$$

this definition yields a Banach space. Moreover, we let

$$\overset{\circ}{W}_F^1(\Omega; \mathbb{R}^N) = \text{the closure of } C_0^\infty(\Omega; \mathbb{R}^N) \text{ in } W_F^1(\Omega; \mathbb{R}^N)$$

$$\text{with respect to } \| \cdot \|_{W_F^1(\Omega;\mathbb{R}^N)} .$$

As proved in [FO], Theorem 2.1, the space $\overset{\circ}{W}_F^1(\Omega; \mathbb{R}^N)$ is characterized by

Theorem 3.1. *Let Ω denote a bounded Lipschitz domain. Then*

$$\overset{\circ}{W}{}^1_F(\Omega; \mathbb{R}^N) = W^1_F(\Omega; \mathbb{R}^N) \cap \overset{\circ}{W}{}^1_1(\Omega; \mathbb{R}^N) .$$

Throughout this chapter F is assumed to satisfy the above hypotheses. Moreover, let us introduce

$$\mathbb{K}_F = u_0 + \overset{\circ}{W}{}^1_F(\Omega; \mathbb{R}^N) , \tag{1}$$

where the Dirichlet boundary data u_0 are assumed to be of class $W^1_F(\Omega; \mathbb{R}^N)$.

With these preliminaries we consider a strictly convex variational integrand f satisfying

$$c_1 F(|Z|) - c_2 \leq f(Z) \quad \text{for all } Z \in \mathbb{R}^{nN} , \tag{2}$$

c_1, c_2 denoting some positive numbers.

Remark 3.2. *Since a corresponding upper bound is not assumed in (2), we introduce in addition to (1) the energy class*

$$\mathbb{J}_F := \left\{ w \in W^1_F(\Omega; \mathbb{R}^N) \text{ such that } \int_\Omega f(\nabla w) \, dx < \infty \right\} .$$

With the above notation we have the following theorem on the existence of unique solutions in Orlicz-Sobolev classes.

Theorem 3.3. *If f is strictly convex and satisfies (2), then the problem*

$$J[w] = \int_\Omega f(\nabla w) \, dx \to \min \quad \text{in the class } \mathbb{K}_F \tag{$\mathcal{P}$}$$

admits a unique solution $u \in \mathbb{K}_F$ provided that we assume $u_0 \in \mathbb{J}_F$.

The *Proof of Theorem 3.3* is given in [FO] under the additional assumption that f is bounded from above by the same N-function. In fact, this restriction is not needed since the energies of minimizing sequences are bounded by the finite value $J[u_0]$.

Let us sketch the main line of the proof: consider a J-minimizing sequence $\{u_m\}$ in $\mathbb{K}_F$, i.e.

$$J[u_m] \overset{m \to \infty}{\Longrightarrow} \gamma = \inf_{\mathbb{K}_F} J > -\infty .$$

On account of $u_0 \in \mathbb{J}_F$, the growth condition (2) implies

$$\int_\Omega F(|\nabla u_m|) \, dx \leq \beta < \infty$$

for some real number $\beta > 0$ which is not depending on m. We may assume that $\beta \geq 1$ which yields by the convexity of F

$$F\left(\frac{1}{\beta}|\nabla u_m|\right) \leq \frac{1}{\beta}F(|\nabla u_m|).$$

By the definition of $\|\cdot\|_{L_F}$, we have proved that $\|\nabla u_m\|_{L_F(\Omega;\mathbb{R}^{nN})} \leq \beta$ and the Poincaré inequality (see Lemma 2.4 of [FO]) gives

$$\|u_m\|_{W_F^1(\Omega;\mathbb{R}^N)} \leq c(\Omega)\left(\beta + \|u_0\|_{W_F^1(\Omega;\mathbb{R}^N)}\right). \tag{3}$$

Next we recall the de la Vallèe Poussin criterion and observe that passing to a subsequence we may assume as $m \to \infty$

$$u_m \rightharpoonup u \quad \text{in } W_1^1\left(\Omega;\mathbb{R}^N\right), \quad u \in u_0 + \overset{\circ}{W}_1^1\left(\Omega;\mathbb{R}^N\right);$$

$$u_m \to u \quad \text{in } L^1\left(\Omega;\mathbb{R}^N\right) \text{ and almost everywhere in } \Omega.$$

Lower semicontinuity with respect to weak W_1^1-convergence then implies $J[u] \leq \gamma$. Hence, u is a suitable candidate to solve $(\mathcal{P})$ and ∇u is of class $L_F(\Omega;\mathbb{R}^{nN})$. It remains to establish $u \in \mathbb{K}_F$: on account of (3) we obtain as above (using in addition (N3))

$$\int_\Omega F(|u_m|) \, dx \leq c,$$

where the constant is not depending on m. Moreover, almost everywhere convergence and Fatou's lemma prove

$$\int_\Omega F(|u|) \, dx < \infty,$$

i.e. $u \in W_F^1(\Omega;\mathbb{R}^N)$ and $u - u_0 \in W_F^1(\Omega;\mathbb{R}^N) \cap \overset{\circ}{W}_1^1(\Omega;\mathbb{R}^N)$. Theorem 3.1 gives $u \in \mathbb{K}_F$, hence, u is a solution of the problem $(\mathcal{P})$. Note that the Δ_2-property (N3) also is needed to establish Theorem 3.1 and the Poincarè inequality of [FO].

The uniqueness of solutions is immediate by the strict convexity of f.   □
The existence and uniqueness theorem is derived by assuming just the growth-condition (2) for the strictly convex integrand f. From now on we concentrate on the regularity theory where we need some additional conditions on the second derivatives of f. These are made precise in the next sections.

3.2 The notion of (s, μ, q)-growth – examples

Definition 3.4. *Let* $F: [0, \infty) \to [0, \infty)$ *denote some N-function satisfying (N1)–(N3). Moreover, fix some real number* $s \geq 1$ *and assume that*

$$F(t) \geq c_0 t^s \quad \text{for large values of } t. \tag{4}$$

A C^2-integrand f is said to be of (s, μ, q)-growth if for all $Z, Y \in \mathbb{R}^{nN}$:

$$c_1 F(|Z|) - c_2 \leq f(Z) ; \tag{5}$$

$$\lambda (1 + |Z|^2)^{-\frac{\mu}{2}} |Y|^2 \leq D^2 f(Z)(Y, Y) \leq \Lambda (1 + |Z|^2)^{\frac{q-2}{2}} |Y|^2 , \tag{6}$$

where $\mu \in \mathbb{R}$, $q > 1$ and c_0, c_1, c_2, λ, Λ denote positive constants.

Remark 3.5. Let us briefly comment on these conditions:

i) Suppose that we are given numbers $q > p > 1$ and that (6) holds with $\mu = 2 - p$. This case corresponds to the version of (p, q)-growth as discussed in A.2 of the introduction.

ii) The logarithmic integrand and its iterated version

$$f(Z) = |Z| \ln \left(1 + \ln \left(1 + \cdots + \ln(1 + |Z|) \ldots \right) \right)$$

are covered by choosing $s = 1$, $\mu = 1$ ($\mu = 1 + \varepsilon$ for the iterated version) and $q = 1 + \varepsilon$ for any $\varepsilon > 0$. Of course the notion of (s, μ, q)-growth also is more general than the preliminary version (3) used in B.2 of the introduction.

iii) Inequality (4) together with the second part of (6) implies $s \leq q$. In fact, consider the case $q < 2$ and observe that it is proved in [AF3], Lemma 2.1: for every $\gamma \in (-1/2, 0)$ and $\mu \geq 0$ we have

$$1 \leq \frac{\int_0^1 \left(\mu^2 + |\eta + s(\zeta - \eta)|^2 \right)^\gamma ds}{\left(\mu^2 + |\zeta|^2 + |\eta|^2 \right)^\gamma} \leq \frac{8}{2\gamma + 1}$$

for all ζ, $\eta \in \mathbb{R}^k$, not both zero if $\mu = 0$. Moreover, we may assume that $f(0) = 0$ and $\nabla f(0) = 0$ (replace f by $f - \nabla f(0) : Z$). This yields

$$\begin{aligned}
f(Z) &= \int_0^1 \int_0^1 D^2 f(t s Z)(Z, Z) \, dt \, s \, ds \\
&\leq |Z|^2 \int_0^1 \int_0^1 \left(1 + t^2 s^2 |Z|^2 \right)^{\frac{q-2}{2}} \, dt \, s \, ds \\
&\leq c |Z|^2 \left(1 + |Z|^2 \right)^{\frac{q-2}{2}} ,
\end{aligned}$$

hence the upper growth rate of f is at most q (in the case $q \geq 2$ this is an obvious consequence of the first inequality without referring to [AF3]).

iv) Since f is smooth, convex and of maximal growth rate q, we also obtain (compare [Da], Lemma 2.2, p. 156)

$$|\nabla f(Z)| \leq c \left(1 + |Z|^2 \right)^{\frac{q-1}{2}}$$

for any $Z \in \mathbb{R}^{nN}$ and for some positive number c – in fact convexity with respect to each z_α^i is sufficient to prove this (see [Ma1], (2.11)).

v) *We may assume that $2 - \mu \leq s$. This is obvious if $\mu \geq 1$ and if $\mu \leq 0$, in the case $0 < \mu < 1$ again compare [AF3], Lemma 2.1.*

In the following we construct an example of an integrand satisfying (4)–(6) precisely with exponents s, μ and q for a given range of values for s, μ and q. Although the construction looks quite technical, only elementary calculations are needed. Note that Example 4.17, iii), below provides some alternative ideas.

We proceed in three steps.

Step 1. Let $q > 1$ and start with the q-power growth integrand $\rho(t) = (1 + t^2)^{q/2}$. Then we "destroy" ellipticity by defining (for all $n \in \mathbb{N}_0$) $\tilde{\rho}(t) = \rho(t)$ if $2n \leq t < 2n + 1$, whereas for $2n + 1 \leq t < 2n + 2$ we let

$$\tilde{\rho}(t) = \rho(2n + 1) + \big(t - (2n + 1)\big)\big(\rho(2n + 2) - \rho(2n + 1)\big) .$$

Moreover, the function $\tilde{\rho}$ is extended to the whole line by setting $\tilde{\rho}(-t) = \tilde{\rho}(t)$.

A mollification $(\tilde{\rho})_\varepsilon$ with some small $\varepsilon > 0$ satisfies:

Lemma 3.6. *There is a positive constant c such that*

i) $(\tilde{\rho})_\varepsilon$ *is a (smooth) N-function.*

ii) *Let $g(Z) = (\tilde{\rho})_\varepsilon(|Z|)$, $Z \in \mathbb{R}^{nN}$. Then we have for all $Z, Y \in \mathbb{R}^{nN}$*

$$0 \leq D^2 g(Z)(Y, Y) \leq c\big(1 + |Z|^2\big)^{\frac{q-2}{2}} |Y|^2 .$$

iii) *The function g satisfies for any $Z \in \mathbb{R}^{nN}$*

$$|D^2 g(Z)||Z|^2 \leq c\big(1 + g(Z)\big) .$$

Proof of Lemma 3.6. By construction we have i). Now fix $\varepsilon = 1/10$ and consider the mollification with kernel k

$$(\tilde{\rho})_\varepsilon(s) = \varepsilon^{-1} \int_{-\infty}^{+\infty} k\left(\frac{s - t}{\varepsilon}\right) \tilde{\rho}(t)\, dt .$$

We fix $n_0 \in \mathbb{N}$ and sketch ii) and iii) for a given $s \in U(t_0)$ where $U(t_0)$ is some small neighborhood of $t_0 = 2n_0 + 1$: to this purpose we let $a = \frac{s - t_0}{\varepsilon}$ and compute

$$(\tilde{\rho})''_\varepsilon(s) = \int_a^\infty k(y)\, \rho''(s - \varepsilon y)\, dy + \frac{k(a)}{\varepsilon}\left(\lim_{t \downarrow t_0} \tilde{\rho}'(t) - \lim_{t \uparrow t_0} \tilde{\rho}'(t)\right) . \quad (7)$$

Now, ρ is strictly convex which implies

$$\lim_{t \downarrow t_0} \tilde{\rho}'(t) \leq \rho'(2n + 2) \quad \text{and} \quad \lim_{t \uparrow t_0} \tilde{\rho}'(t) \geq \rho'(2n) ,$$

and by (7) there is a constant (depending on ε) such that

$$(\tilde{\rho})''_\varepsilon(s) \le (\rho)''_\varepsilon(s) + c\left(\rho'(2n+2) - \rho'(2n)\right)$$

$$= (\rho)''_\varepsilon(s) + c\,\rho''(\xi)\,, \quad \xi \in (2n, 2n+2)\,. \tag{8}$$

Inequality (8) provides the lemma with some additional direct computations.
$\square$

Step 2. For the construction of a degenerate anisotropic (s, q)-power growth integrand let us assume for simplicity that $n = 3$ and $N = 1$. Suppose first that we are given numbers $s \le q$, $2 \le q$. We let

$$h(Z) = (\tilde{\rho}_s)_\varepsilon\left(|z_1|\right) + (\tilde{\rho}_q)_\varepsilon\left(\sqrt{z_2^2 + z_3^2}\right)\,,$$

where $(\tilde{\rho}_s)_\varepsilon$ and $(\tilde{\rho}_q)_\varepsilon$ are defined as above with respect to the exponents s and q.

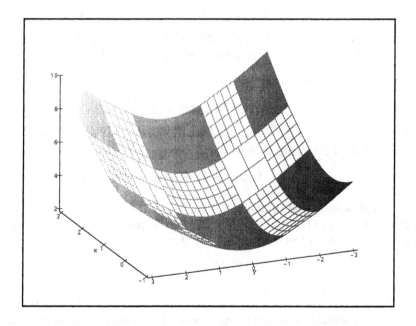

Fig. 1. (s, q)-growth with "linear pieces"

We now have

$$c\left(1 + |Z|^2\right)^{\frac{s}{2}} \le h(Z) \le C\left(1 + |Z|^2\right)^{\frac{q}{2}}\,, \tag{9}$$

$$0 \le D^2 h(Z)(Y, Y) \le C\left(1 + |Z|^2\right)^{\frac{q-2}{2}}|Y|^2 \tag{10}$$

for all $Z, Y \in \mathbb{R}^3$ with positive constants c and C. Note that the exponents in (9) and (10) cannot be improved, moreover, due to the degeneracy of $D^2 h$, the lower bound in (10) is the best possible one.

In the case $1 < q < 2$ the right-hand side inequality of (10) fails to be true. Here the example is modified by letting

$$h(Z) = (\tilde{\rho}_s)_\varepsilon(\Gamma) , \quad \Gamma := \left(z_1^2 + z_2^2 + |z_3|^{2(1+\gamma)} \right)^{\frac{1}{2}}$$

for some appropriate $\gamma > 0$. Then (10) (for some $q > s$) and the first inequality of (9) are valid, the second one holds for some $\tilde{q}$, $s < \tilde{q} < q$.

Step 3. In the last step let us fix a μ-elliptic function Φ, i.e. Φ is assumed to satisfy the left-hand inequality of (6). Moreover, we suppose that the upper growth rate of Φ is less than or equal to s (compare Remark 3.8 and Example 3.9 for particular choices). As a result we obtain

Example 3.7. *Suppose that we are given numbers μ, s and q such that*

$$(2 - \mu) \leq s \leq q .$$

With the above notation let $f(Z) = h(Z) + \Phi(Z)$. Then f is an integrand of (s, μ, q)-growth such that the exponents cannot be improved.

Remark 3.8.   *i) In the case*

$$1 \leq p := 2 - \mu \leq s \leq q ,$$

Φ *may be chosen as the power growth function* $\Phi(Z) = (1 + |Z|^2)^{p/2}$.
ii) Non-standard ellipticity of Φ also is possible:

$$\lambda \left(1 + |Z|^2 \right)^{-\frac{\mu}{2}} |Y|^2 \leq D^2 \Phi(Z)(Y, Y) , \quad \mu > 1 .$$

This will be discussed in Example 3.9.
iii) If we choose $f(Z) = g(Z) + \Phi(Z)$ where g is given in Lemma 3.6, then $s = q$ and f is "balanced" in the sense that

$$|D^2 f(Z)| |Z|^2 \leq c \left(1 + f(Z) \right)$$

(compare Lemma 3.6, iii)).

We finish this section by constructing a class of non-standard μ-elliptic integrands, i.e. the ellipticity condition holds with some exponent $\mu > 1$.

Example 3.9. *Given $\mu > 1$ let*

$$\varphi(r) = \int_0^r \int_0^s \left(1 + t^2 \right)^{-\frac{\mu}{2}} dt \, ds , \qquad\qquad r \in \mathbb{R}_0^+ ,$$

$$\Phi(Z) = \int_0^{|Z|} \int_0^s \left(1 + t^2 \right)^{-\frac{\mu}{2}} dt \, ds = \varphi(|Z|) , \quad Z \in \mathbb{R}^{nN} .$$

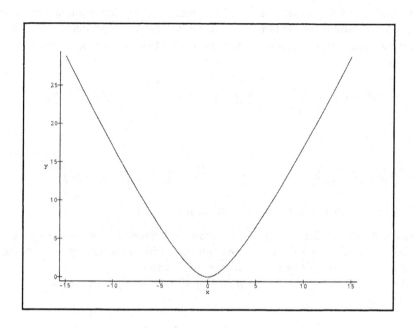

Fig. 2. An integrand of linear growth with μ-ellipticity, $\mu = 4/3$.

The function Φ is of linear growth and satisfies for all Z, $Y \in \mathbb{R}^{nN}$ with suitable constants c, $C > 0$

i) $$D\Phi(Z) = Z \int_0^1 \left(1 + t^2 |Z|^2\right)^{-\frac{\mu}{2}} dt \; ;$$

ii) $$\frac{\partial^2 \Phi}{\partial z_\alpha^i \partial z_\beta^j}(Z) = \left[\delta^{ij}\, \delta_{\alpha\beta} - |Z|^{-2} z_\alpha^i z_\beta^j\right] \int_0^1 \left(1 + t^2 |Z|^2\right)^{-\frac{\mu}{2}} dt$$
$$+ |Z|^{-2} z_\alpha^i z_\beta^j \left(1 + |Z|^2\right)^{-\frac{\mu}{2}} \; ;$$

iii) $c\left(1 + |Z|^2\right)^{-\frac{\mu}{2}} |Y|^2 \leq D^2\Phi(Z)(Y,Y) \leq C\left(1 + |Z|^2\right)^{-\frac{1}{2}} |Y|^2;$

iv) $$|D^2\Phi(Z)|\,|Z|^2 \leq C|Z| \; .$$

Proof of the above stated properties. The choice of μ shows that $\varphi'(r) \leq \int_0^\infty (1 + t^2)^{-\mu/2}\, dt < \infty$. Hence it follows that Φ is at most of linear growth. The fact that Φ is at least of linear growth is immediate by definition. Note that this in general is true for μ-elliptic integrands (compare Remark 4.2, *ii*), below).

Using a linear transformation, the proof of *i*) and *ii*) is obvious. Moreover, the first inequality stated in *iii*) is a consequence of *ii*) and follows by considering the cases $|Y : Z| \leq \frac{1}{2}|Y||Z|$ and $|Y : Z| > \frac{1}{2}|Y||Z|$, respectively. We like to remark that the exponent $-\mu/2$ occurring on the left-hand side of *iii)* is the best possible one. This is evident if we consider Y parallel to Z. The second inequality of *iii)* again is immediate. Next we are going to prove *iv)*: observing

$$|D^2 \Phi(Z)| = \sup_{|Y|=1} D^2 \Phi(Z)(Y, Y) \leq 2 \int_0^1 \left(1 + t^2 |Z|^2\right)^{-\frac{\mu}{2}} dt$$

we get

$$|D^2 \Phi(Z)| |Z|^2 \leq 2|Z| \int_0^{|Z|} \left(1 + s^2\right)^{-\frac{\mu}{2}} ds \leq 2|Z| \int_0^\infty \left(1 + s^2\right)^{-\frac{\mu}{2}} ds ,$$

the last integral being finite on account of $\mu > 1$. □

Remark 3.10. *i) Let us already mention that Example 3.9 provides a regular class of integrands with linear growth (with some appropriate choice of $\mu > 1$). This will be proved in the next chapter.*

ii) We refer to Example 4.17 for a further discussion of Φ in the case $\mu \geq 1$.

3.3 A priori gradient bounds and local $C^{1,\alpha}$-estimates for scalar and structured vector-valued problems

We are going to prove the Theorems 1 and 3 as stated in the introduction for our new class of variational integrals of (s, μ, q)-type. Let us assume in the vectorial setting $N > 1$ that f is of "special structure" in the sense that

$$f(Z) = g(|Z|^2) , \qquad Z \in \mathbb{R}^{nN} , \tag{11}$$

holds with $g: [0, \infty) \to [0, \infty)$ of class C^2. Note that this implies

$$\frac{\partial^2 f}{\partial z_\alpha^i \, \partial z_\beta^j}(Z) = 4g''(|Z|^2) z_\alpha^i z_\beta^j + 2g'(|Z|^2) \delta^{ij} \delta_{\alpha\beta} , \tag{12}$$

$\alpha, \beta = 1, \ldots, n$; $i, j = 1, \ldots, N$. We also assume in the case $N > 1$ that there are real numbers $\alpha \in (0, 1]$, $K > 0$ satisfying for all $Z, \tilde{Z} \in \mathbb{R}^{nN}$

$$\left|D^2 f(Z) - D^2 f(\tilde{Z})\right| \leq K |Z - \tilde{Z}|^\alpha . \tag{13}$$

Note that the above examples are easily adjusted to this conditions (in fact to much stronger ones) by letting

$$\varphi(r) = \int_0^{\sqrt{\varepsilon + r^2}} \int_0^s \left(1 + t^2\right)^{-\frac{\mu}{2}} dt \, ds , \qquad \varepsilon > 0 .$$

Theorem 3.11. *Given $\mu \in (-\infty, 2)$, $1 \le s \le q$, $1 < q$, assume that f is of (s, μ, q)-growth. In the vectorial case suppose in addition that (11) and (13) are satisfied. Finally let u denote the unique solution of the problem*

$$J[w] = \int_\Omega f(\nabla w) \, dx \to \min \quad in \ \mathbb{K}_F \qquad (\mathcal{P})$$

with boundary values $u_0 \in \mathbb{J}_F$.

i) If the (s, μ, q)-condition

$$q < (2 - \mu) + s \frac{2}{n} \qquad (14)$$

holds, then u is of class $C^{1,\alpha}(\Omega; \mathbb{R}^N)$ for any $0 < \alpha < 1$.
ii) If f is "balanced" in the sense that (compare Remark 3.8)

$$|D^2 f(Z)||Z|^2 \le c(1 + f(Z)) \qquad (15)$$

holds for all $Z \in \mathbb{R}^{nN}$ and for some real number $c > 0$, then we may replace (14) by the weaker condition

$$q < (2 - \mu) \frac{n}{n - 2} \quad if \ n > 2. \qquad (16)$$

Remark 3.12.

i) Note that on account of $q > 1$ the condition (16) gives the upper bound

$$\mu < 1 + \frac{2}{n} \quad if \ n > 2.$$

Our choice $\mu < 2$ however implies that the latter inequality is valid in all dimensions.

ii) Theorem 3.11 is proved in [BFM], Theorem 1.1, for the scalar case $N = 1$, where the assertion is formulated for local minimizers. Additionally, the results are shown to be true for (double)-obstacle problems with minimal assumptions on the obstacles. Since these aspects can be included with slight modifications of the proof given below, we keep the main line and consider unconstrained Dirichlet problems.

iii) In the vectorial setting we benefit, for instance, from [FuM] and [BF3] (compare [Uh], [GiaM]). It should be mentioned that a priori gradient bounds can be derived without condition (13) which is only needed for proving local $C^{1,\alpha}$-regularity.

Proof of Theorem 3.11. The proof of Theorem 3.11 is organized in four steps: regularization, a priori L^q-estimates, a priori L^∞-estimates and the conclusion.

3.3.1 Regularization

In the following we fix a bounded Lipschitz domain $\hat{\Omega}$ and boundary values $\hat{u}_0 \in W_q^1(\hat{\Omega}; \mathbb{R}^N)$. The data Ω and u_0 of Theorem 3.11 are recovered in Section 3.3.4. The regularization $u_\delta \in \hat{u}_0 + \overset{\circ}{W}_q^1(\hat{\Omega}; \mathbb{R}^N)$ is defined as the unique solution of the Dirichlet problem

$$J_\delta[w] = \int_{\hat{\Omega}} f_\delta(\nabla w) \, dx \to \min \quad \text{in } \hat{u}_0 + \overset{\circ}{W}_q^1(\hat{\Omega}; \mathbb{R}^N),$$

where for any $0 < \delta < 1$

$$f_\delta(Z) := f(Z) + \delta \left(1 + |Z|^2\right)^{\frac{q}{2}} .$$

Remark 3.13. *Note that the regularization is done with respect to q. Analogous to Section 2.1.2 we could try to add a δ-term of power $t > q$ instead of the above choice. This, however, would yield some serious technical difficulties in the scalar case (and for vectorial problems with additional structure), where it is not evident how to handle higher order δ-terms in order to make the DeGiorgi technique work.*

In the vector-valued setting of Section 3.4, the choice $t = q$ gives some difficulties concerning the starting integrability since we cannot refer to the discussion of asymptotic regular integrands as given in [CE] or [GiaM]. On the other hand, once the Caccioppoli-type inequality of Lemma 3.19 is established, the absence of additional δ-terms simplifies the proof of the uniform local higher integrability Lemma 3.17. This is why we use the above approach throughout the whole Chapter.

Analogous to Section 2.1.2 we have the Euler equation

$$\int_{\hat{\Omega}} \nabla f_\delta(\nabla u_\delta) : \nabla \varphi \, dx = 0 \quad \text{for all } \varphi \in \overset{\circ}{W}_q^1(\hat{\Omega}; \mathbb{R}^N) \tag{17}$$

together with the uniform bound

$$J_\delta[u_\delta] \leq J_\delta[\hat{u}_0] = \int_{\hat{\Omega}} f_\delta(\nabla \hat{u}_0) \, dx . \tag{18}$$

Studying the properties of u_δ let us first consider the scalar case. Then, from standard arguments (see, e.g. [LU1], Chapter 4, Theorem 5.2, p. 277) we get $u_\delta \in W_{2,loc}^2(\hat{\Omega}) \cap W_{\infty,loc}^1(\hat{\Omega})$ using also the fact that u_δ is locally bounded which is proved under very weak assumptions in [GG]. Now fix a subdomain $\Omega' \Subset \hat{\Omega}$ and a coordinate direction $\gamma \in \{1, \ldots, n\}$. From (17) we obtain the differentiated Euler equation

$$\int_{\Omega'} D^2 f(\nabla u_\delta)(\nabla \partial_\gamma u_\delta, \nabla \varphi) \, dx = 0 \quad \text{for all } \varphi \in C_0^1(\Omega') ,$$

where the coefficients $\frac{\partial^2 f}{\partial z_\alpha \partial z_\beta}(\nabla u_\delta)$ of class $C^0(\overline{\Omega'})$ are uniformly elliptic. Quoting "L^p-theory" for equations with continuous coefficients (see [Mor1], Theorem 5.5.3, p. 154, or [Gia3], Chapter 4.3, pp. 71) we get $\partial_\gamma u_\delta \in W^1_{t,loc}(\Omega')$ for any finite t. Thus we have $u_\delta \in W^2_{t,loc}(\Omega)$ for any $1 \le t < \infty$.

In the vectorial case, some well known properties of u_δ are summarized in the following lemma. Part I) is proved in [AF3], Proposition 2.4 and Lemma 2.5, for the second part we refer the reader to [GiaM], especially formula (3.3), and to [Ca] (compare Theorem 1.1).

Lemma 3.14.

I) *In the case $q < 2$ the approximative solution satisfies:*

$i)$ $u_\delta \in W^2_{q,loc}(\hat{\Omega}; \mathbb{R}^N)$;

$ii)$ $\nabla f_\delta(\nabla u_\delta) \in W^1_{2,loc}(\hat{\Omega}; \mathbb{R}^{nN})$;

$iii)$ $\left(1 + |\nabla u_\delta|^2\right)^{\frac{q-2}{4}} \nabla u_\delta \in W^1_{2,loc}(\hat{\Omega}; \mathbb{R}^{nN})$;

$iv)$ $\left|\nabla^2 u_\delta \mathbf{1}_{[|\nabla u_\delta| \le M]}\right| \in L^2_{loc}(\hat{\Omega})$ *for all $M > 0$.*

II) *In the case $q \ge 2$ we have*

$i)$ $u_\delta \in W^2_{2,loc}(\hat{\Omega}; \mathbb{R}^N)$;

$ii)$ $\nabla f_\delta(\nabla u_\delta) \in W^1_{q/(q-1),loc}(\hat{\Omega}; \mathbb{R}^{nN})$;

$iii)$ $\left(1 + |\nabla u_\delta|^2\right)^{\frac{q}{4}} \in W^1_{2,loc}(\hat{\Omega})$;

$iv)$ $\left(1 + |\nabla u_\delta|^2\right)^{\frac{q-2}{4}} \nabla u_\delta \in W^1_{2,loc}(\hat{\Omega}; \mathbb{R}^{nN})$.

Remark 3.15. *On account of the structure condition (11) the properties of u_δ are "better" than stated in Lemma 3.14 (see, for instance, [AF3], Proposition 2.7). This information is not used for the moment, since uniform L^q-estimates do not depend on this special structure.*

In Section 3.3.4, Theorem 3.11 will be proved as a corollary of the uniform a priori L^∞-estimates formulated in

Theorem 3.16. *Consider a ball $B_{2R_0}(x_0) \Subset \hat{\Omega}$ and suppose that f satisfies the assumptions of Theorem 3.11. Then there is a real number*

$$c = c\left(J_\delta[u_\delta], R_0\right),$$

depending in addition only on n, N and the data of f, such that

$$\|\nabla u_\delta\|_{L^\infty(B_{R_0/2}(x_0); \mathbb{R}^{nN})} \le c.$$

This Theorem will be proved in Section 3.3.3.

In the following we fix $x_0 \in \hat{\Omega}$ and $R_0 > 0$ according to the above condition $B_{2R_0}(x_0) \Subset \hat{\Omega}$. Moreover, we introduce the notation $c = c(f)$ for constants c depending only on n, N and the data of the variational integrand f.

3.3.2 A priori L^q-estimates

The main part of the second step is to prove the following lemma which also holds true in the case of vectorial problems without additional structure (compare Remark 3.15). As a consequence, Lemma 3.17 will also serve as an important tool while proving the partial regularity results of the next chapter.

In addition to R_0 we now fix radii r, R such that $0 < r < 2R \leq 2R_0$.

Lemma 3.17. *Let f satisfy the assumptions of Theorem 3.11 without the restrictions (11) and (13). Moreover, let $\chi := n/(n-2)$, if $n \geq 3$. In the case $n = 2$ define a number $\chi > 1$ through the condition*

$$\chi \begin{cases} > \dfrac{s}{s + 2 - \mu - q} & \text{in the case i) of Theorem 3.11;} \\[2ex] > \dfrac{q}{2 - \mu} & \text{in the case ii) of Theorem 3.11.} \end{cases}$$

Then there are constants $c \equiv c(f, r, R)$, $\beta \equiv \beta(f)$, independent of δ, such that:

$$\int_{B_r(x_0)} \left(1 + |\nabla u_\delta|^2\right)^{\frac{(2-\mu)\chi}{2}} dx \leq c \left\{ \int_{B_{2R}(x_0)} \left(1 + f_\delta(\nabla u_\delta)\right) dx \right\}^\beta.$$

Remark 3.18. *Note that our assumptions imply $q < (2 - \mu)\chi$. The proof given below in fact will show that in the two-dimensional case $n = 2$ we can choose χ as any finite number. Of course the constants will then depend on the quantity χ.*

The starting point for a proof of Lemma 3.17 is the following Caccioppoli-type inequality for the approximative solutions.

Lemma 3.19. *There is a real number $c > 0$ such that for all $\eta \in C_0^1(B_{2R}(x_0))$, $0 \leq \eta \leq 1$, and for all $Q \in \mathbb{R}^{nN}$*

$$\int_{B_{2R}(x_0)} \eta^2 \, D^2 f_\delta(\nabla u_\delta) \left(\partial_\gamma \nabla u_\delta, \partial_\gamma \nabla u_\delta\right) dx$$

$$\leq c \|\nabla \eta\|_\infty^2 \int_{\text{spt} \nabla \eta} \left|D^2 f_\delta(\nabla u_\delta)\right| |\nabla u_\delta - Q|^2 \, dx \,,$$

where the summation with respect to $\gamma = 1, \ldots, n$ is always assumed. In particular, for all $Q \in \mathbb{R}^{nN}$

$$\int_{B_{2R}(x_0)} \eta^2 \left(1 + |\nabla u_\delta|^2\right)^{-\frac{\mu}{2}} |\nabla^2 u_\delta|^2 \, dx$$

$$\leq c \|\nabla \eta\|_\infty^2 \int_{\mathrm{spt}\,\nabla\eta} |D^2 f_\delta(\nabla u_\delta)| \, |\nabla u_\delta - Q|^2 \, dx \, .$$

Proof of Lemma 3.19. We first recall that u_δ solves the regularized problem, i.e. we have (17). Next denote by $e_\gamma \in \mathbb{R}^n$ the unit coordinate vector in x_γ-direction and let for a function g on $\hat\Omega$

$$\Delta_h g(x) = \Delta_h^\gamma g(x) = \frac{g(x + h e_\gamma) - g(x)}{h} \, , \quad h \in \mathbb{R} \, ,$$

denote the difference quotient of g at x in the direction e_γ. Then, given $Q \in \mathbb{R}^{nN}$, $\eta \in C_0^1(B_{2R}(x_0))$, the choice $\varphi = \Delta_{-h}(\eta^2 \Delta_h (u_\delta - Qx))$, is admissible in (17) and an "integration by parts" implies

$$\int_{B_{2R}(x_0)} \eta^2 \Delta_h\big(\nabla f_\delta(\nabla u_\delta)\big) : \nabla \Delta_h u_\delta \, dx$$

$$= -2 \int_{B_{2R}(x_0)} \eta \Delta_h\big(\nabla f_\delta(\nabla u_\delta)\big) : \Delta_h(u_\delta - Qx) \otimes \nabla\eta \, dx \, . \tag{19}$$

We start with the consideration of the case $q \geq 2$: by Lemma 3.14, ∇u_δ is known to be of class L_{loc}^r for some $r > q$ and if F_h denotes the integrand on the right-hand side of (19), then the existence of a real number $c(\nabla\eta)$, independent of h, follows such that

$$|F_h| \leq c \left\{ \left|\Delta_h\big(\nabla f_\delta(\nabla u_\delta)\big)\right|^{l_1} + |\Delta_h u_\delta|^{l_2} \right\} \text{ for some } l_1 < \frac{q}{q-1} \, , \quad q < l_2 < r.$$

Thus, equiintegrability of F_h in the sense of Vitali's convergence theorem is established by Lemma 3.14, II), *ii*) and, passing to the limit $h \to 0$, the right-hand side of (19) tends to

$$-2 \int_{B_{2R}(x_0)} \eta \partial_\gamma\big(\nabla f_\delta(\nabla u_\delta)\big) : (\partial_\gamma u_\delta - Q_\gamma) \otimes \nabla\eta \, dx \in (-\infty, +\infty) \, . \tag{20}$$

For the left-hand side of (19) we observe

$$\Delta_h\big(\nabla f_\delta(\nabla u_\delta)\big) = \int_0^1 D^2 f_\delta\big(\nabla u_\delta + t h \Delta_h \nabla u_\delta\big)\big(\Delta_h \nabla u_\delta, \cdot\big) \, dt$$

and get using (20), Fatou's lemma and Young's inequality

$$\int_{B_{2R}(x_0)} \eta^2 D^2 f_\delta(\nabla u_\delta)(\partial_\gamma \nabla u_\delta, \partial_\gamma \nabla u_\delta) \, dx$$

$$\leq \int_{B_{2R}(x_0)} \eta^2 \liminf_{h \to 0} \int_0^1 D^2 f_\delta(\nabla u_\delta + t h \Delta_h \nabla u_\delta)(\Delta_h \nabla u_\delta, \Delta_h \nabla u_\delta) \, dt \, dx$$

$$\leq \frac{1}{2} \int_{B_{2R}(x_0)} \eta^2 D^2 f_\delta(\nabla u_\delta)(\partial_\gamma \nabla u_\delta, \partial_\gamma \nabla u_\delta) \, dx$$

$$+ c \|\nabla \eta\|_\infty^2 \int_{\mathrm{spt} \nabla \eta} |D^2 f_\delta(\nabla u_\delta)| |\nabla u_\delta - Q|^2 \, dx \, ,$$

i.e. the lemma is proved for $q \geq 2$.

In the case $q < 2$ we modify the truncation arguments of [EM1]. To this purpose we fix $M \gg 1$ and let for $t \geq 0$

$$\psi(t) := \begin{cases} 0 \, , t \geq M \\ 1 \, , t \leq M/2 \end{cases} \, , \quad |\psi'(t)| \leq 4/M \, .$$

Now, by Lemma 3.14, I), iv), and by [EM1], Lemma 1, $\varphi = \Delta_{-h}\big(\eta^2 \, \partial_\gamma(u_\delta - Qx)\psi(|\nabla u_\delta|)\big)$ is an admissible choice, hence

$$\int_{B_{2R}(x_0)} \eta^2 \psi \Delta_h\big(\nabla f_\delta(\nabla u_\delta)\big) : \nabla \partial_\gamma u_\delta \, dx$$

$$= -2 \int_{B_{2R}(x_0)} \eta \psi \Delta_h\big(\nabla f_\delta(\nabla u_\delta)\big) : \partial_\gamma(u_\delta - Qx) \otimes \nabla \eta \, dx \qquad (21)$$

$$- 2 \int_{B_{2R}(x_0)} \eta^2 \Delta_h\big(\nabla f_\delta(\nabla u_\delta)\big) : \partial_\gamma(u_\delta - Qx) \otimes \nabla \psi \, dx \, .$$

By the definition of ψ and again on account of Lemma 3.14, I), iv), both integrals on the right-hand side of (21) can be written as

$$\int_{\mathrm{spt} \, \eta} \Delta_h\big(\nabla f_\delta(\nabla u_\delta)\big) : \xi(x) \, dx$$

$$\leq \int_{\mathrm{spt} \, \eta} \big|\Delta_h\big(\nabla f_\delta(\nabla u_\delta)\big)\big|^2 \, dx + \|\xi\|^2_{L^2(B_{2R}(x_0); \mathbb{R}^{nN})} \, , \qquad (22)$$

where ξ is a suitable function of class L^2. Since Lemma 3.14, I), ii), shows that $\partial_\gamma\big(\nabla f_\delta(\nabla u_\delta)\big)$ is of class L^2_{loc}, the strong convergence of difference quotients (see [Mor1], Theorem 3.6.8 (b), p. 84) implies (passing to the limit $h \to 0$)

$$\big|\Delta_h\big(\nabla f_\delta(\nabla u_\delta)\big)\big|^2 \to \big|\partial_\gamma\big(\nabla f_\delta(\nabla u_\delta)\big)\big|^2 \quad \text{a.e.} \, ,$$

$$\int_{\mathrm{spt} \, \eta} \big|\Delta_h\big(\nabla f_\delta(\nabla u_\delta)\big)\big|^2 \, dx \to \int_{\mathrm{spt} \, \eta} \big|\partial_\gamma\big(\nabla f_\delta(\nabla u_\delta)\big)\big|^2 \, dx \, . \qquad (23)$$

With (22) and (23) the variant of the Dominated Convergence Theorem given for example in [EG2], Theorem 4, p. 21, is applicable (note that the almost everywhere convergence in (23) is needed for a proof of this variant). Thus, we may pass to the limit $h \to 0$ on the right-hand side of (21). The left-hand side is handled as in the case $q \geq 2$ and summarizing the results we arrive at (again after applying Young's inequality to the bilinear form $D^2 f_\delta(\nabla u_\delta)$)

$$\int_{B_{2R}(x_0)} \eta^2 \psi D^2 f_\delta(\nabla u_\delta)(\partial_\gamma \nabla u_\delta, \partial_\gamma \nabla u_\delta) \, \mathrm{d}x$$

$$\leq \frac{1}{2} \int_{B_{2R}(x_0)} \eta^2 \psi D^2 f_\delta(\nabla u_\delta)(\partial_\gamma \nabla u_\delta, \partial_\gamma \nabla u_\delta) \, \mathrm{d}x$$

$$+ c \|\nabla \eta\|_\infty^2 \int_{\mathrm{spt}\, \nabla \eta} |D^2 f_\delta(\nabla u_\delta)| |\nabla u_\delta - Q|^2 \, \mathrm{d}x$$

$$+ c \int_{\mathrm{spt}\, \eta} |D^2 f_\delta(\nabla u_\delta)| |\nabla^2 u_\delta|^2 \, \mathbf{1}_{[M/2 \leq |\nabla u_\delta| \leq M]} \, \mathrm{d}x \,.$$

To get this inequality we note in addition that $|\nabla u_\delta - Q| < 2M$ on $[M/2 \leq |\nabla u_\delta| \leq M]$ for M sufficiently large and that $\nabla(\psi(|\nabla u_\delta|)) \leq c |\nabla^2 u_\delta|/M$. Before passing to the limit $M \to \infty$, we use Proposition 2.4 of [AF3] once again, i.e. we observe the estimate

$$\int_{B_t(x_0)} \left(1 + |\nabla u_\delta|^2\right)^{\frac{q-2}{2}} |\nabla^2 u_\delta|^2 \, \mathrm{d}x \leq c(t,t') \int_{B_{t'}(x_0)} \left(1 + |\nabla u_\delta|^2\right)^{\frac{q}{2}} \, \mathrm{d}x$$

being valid for all $0 < t < t' < 2R$. Recalling the growth of $|D^2 f_\delta(\nabla u_\delta)|$ we then immediately get

$$\int_{\mathrm{spt}\, \eta} |D^2 f_\delta(\nabla u_\delta)| |\nabla^2 u_\delta|^2 \, \mathbf{1}_{[M/2 \leq |\nabla u_\delta| \leq M]} \, \mathrm{d}x \xrightarrow{M \to \infty} 0$$

on account of $\mathbf{1}_{\mathrm{spt}\, \nabla \psi} \to 0$ as $M \to \infty$ and the claim of the lemma follows.
$\square$

   Besides Lemma 3.19, the following technical proposition is needed to prove uniform L^q-estimates for ∇u_δ. So let us introduce $\Theta(t) := (1 + t^2)^{(2-\mu)/4}$, $t \geq 0$, and let $h_\delta := \Theta(|\nabla u_\delta|)$.

Proposition 3.20. *With this notation we have* $h_\delta \in W^1_{2,loc}(B_{2R}(x_0))$ *and*

$$\nabla h_\delta = \Theta'(|\nabla u_\delta|) \nabla |\nabla u_\delta| \,.$$

Remark 3.21. *If we consider, for instance, the case $q \geq 2$, then the fact that h_δ is of class $W^1_{2,loc}$ follows from Lemma 3.14, II), iii). However, proving Lemma 3.17 we need an explicit formula for the derivative.*

Proof of Proposition 3.20. In order to reduce the problem to an application of the usual chain rule for Lipschitz functions, let $L \gg 1$ be some real number and let

$$\Theta_L(t) = \begin{cases} (1+t^2)^{\frac{2-\mu}{4}}, & 0 \leq t \leq L, \\ (1+L^2)^{\frac{2-\mu}{4}}, & t \geq L, \end{cases} \quad ; \quad h_\delta^L := \Theta_L(|\nabla u_\delta|).$$

As a consequence of Lemma 3.14, h_δ^L is immediately seen to be of class $W_{1,loc}^1$ satisfying

$$\nabla h_\delta^L = \Theta_L'(|\nabla u_\delta|) \nabla |\nabla u_\delta|. \tag{24}$$

In addition, for $0 < r < 2R$ we have the estimate

$$\int_{B_r(x_0)} |\nabla h_\delta^L|^2 \, dx \leq \int_{B_r(x_0) \cap [|\nabla u_\delta| \leq L]} |\Theta'|^2 |\nabla^2 u_\delta|^2 \, dx$$
$$\leq c \int_{B_{2R}(x_0)} (1 + |\nabla u_\delta|^2)^{-\frac{\mu}{2}} |\nabla^2 u_\delta|^2 \, dx,$$

hence, by Lemma 3.19, $\|\nabla h_\delta^L\|_{L^2(B_r(x_0), \mathbb{R}^n)}$ is uniformly bounded with respect to L and we may assume

$$\nabla h_\delta^L \rightharpoonup: W_\delta \quad \text{in } L^2(B_r(x_0), \mathbb{R}^n) \text{ as } L \to \infty.$$

On the other hand, the obvious convergence $h_\delta^L \to h_\delta$ in $L^2(B_r(x_0))$ as $L \to \infty$ implies $W_\delta = \nabla h_\delta$, thus $h_\delta \in W_2^1(B_r(x_0))$. Moreover, (24) gives

$$\nabla h_\delta^L \to \Theta'(|\nabla u_\delta|) \nabla |\nabla u_\delta| \quad \text{almost everywhere},$$

hence we can identify the limit and the proposition is proved. $\square$

Now we are going to *prove Lemma 3.17*: starting with the case $n \geq 3$, we let $\alpha = \frac{(2-\mu)n}{2(n-2)}$ and assume without loss of generality $R < r < 3R/2$. Moreover, fix $0 < \rho < R/2$ and $\eta \in C_0^1(B_{r+\rho/2}(x_0))$, $\eta \equiv 1$ on $B_r(x_0)$, $\nabla \eta \leq 4/\rho$. Since h_δ is of class W_2^1 (see Proposition 3.20), we obtain using Sobolev's inequality

$$\int_{B_r(x_0)} (1 + |\nabla u_\delta|^2)^\alpha \, dx \leq \int_{B_{2R}(x_0)} \left(\eta \left[1 + |\nabla u_\delta|^2 \right]^{\alpha \frac{n-2}{2n}} \right)^{\frac{2n}{n-2}} \, dx$$
$$= \int_{B_{2R}(x_0)} (\eta h_\delta)^{\frac{2n}{n-2}} \, dx$$
$$\leq c \left(\int_{B_{2R}(x_0)} |\nabla(\eta h_\delta)|^2 \, dx \right)^{\frac{n}{n-2}}$$
$$\leq c(n) \{ T_1 + T_2 \}^{\frac{n}{n-2}},$$

where we have set

$$T_1 = \int_{B_{2R}(x_0)} |\nabla \eta|^2 h_\delta^2 \, dx, \quad T_2 = \int_{B_{2R}(x_0)} \eta^2 |\nabla h_\delta|^2 \, dx.$$

T_1 satisfies

$$T_1 \le \frac{c}{\rho^2} \int_{B_{2R}(x_0)} \left(1 + |\nabla u_\delta|^2\right)^{\frac{2-\mu}{2}} dx \,,$$

whereas T_2 has to be handled via the representation formula for the derivative of h_δ given in Proposition 3.20:

$$T_2 \le c \int_{B_{r+\rho/2}(x_0)} \left(1 + |\nabla u_\delta|^2\right)^{-\frac{\mu}{2}} |\nabla^2 u_\delta|^2 \, dx \,.$$

With Lemma 3.19 (choosing $Q = 0$) and (6) of Section 3.2 we immediately have Lemma 3.17 in the case that the assumptions ii) of Theorem 3.11 are valid. If i) is assumed, then we obtain (once more using Lemma 3.19 and (6), Section 3.2)

$$\left(\int_{B_r(x_0)} \left(1 + |\nabla u_\delta|^2\right)^\alpha dx\right)^{\frac{1}{\chi}} \le \frac{c}{\rho^2} \left\{\int_{B_{2R}(x_0)} \left(1 + |\nabla u_\delta|^2\right)^{\frac{2-\mu}{2}} dx \right.$$
$$\left. + \int_{(B_{r+\rho}-B_r)(x_0)} \left(1 + |\nabla u_\delta|^2\right)^{\frac{q}{2}} dx\right\} . \quad (25)$$

Now, observe that inequality (25) is more or less the same one as given after (4.6) of [ELM1], hence we proceed with the ideas outlined there: choose $\theta \in (0,1)$ such that

$$\frac{1}{q} = \frac{\theta}{s} + \frac{1-\theta}{(2-\mu)\chi} \,.$$

Then the well known interpolation inequality (compare [GT], (7.9), p. 146) implies

$$\|\nabla u_\delta\|_q \le \|\nabla u_\delta\|_s^\theta \|\nabla u_\delta\|_{(2-\mu)\chi}^{1-\theta} \,,$$

where the norms are taken over $T_{r,\rho} := B_{r+\rho}(x_0) - B_r(x_0)$. We obtain

$$\frac{1}{\rho^2} \int_{T_{r,\rho}} |\nabla u_\delta|^q \, dx \le \frac{1}{\rho^2} \left(\int_{B_{2R}(x_0)} |\nabla u_\delta|^s \, dx\right)^{\frac{\theta q}{s}}$$
$$\cdot \left(\int_{T_{r,\rho}} |\nabla u_\delta|^{(2-\mu)\chi} \, dx\right)^{\frac{(1-\theta)q}{(2-\mu)\chi}} \,.$$

The choice of θ together with the (s,μ,q)-condition (14) yields $(1-\theta)q/(2-\mu) < 1$, thus there exist some positive numbers β_1, β_2 such that

$$\frac{1}{\rho^2} \int_{T_{r,\rho}} |\nabla u_\delta|^q \, dx \le \frac{1}{\rho^{\beta_1}} \left(\int_{B_{2R}(x_0)} |\nabla u_\delta|^s \, dx\right)^{\beta_2}$$
$$+ \left(\int_{T_{r,\rho}} |\nabla u_\delta|^{(2-\mu)\chi} \, dx\right)^{\frac{1}{\chi}} \,. \quad (26)$$

Combining (25) and (26) we recall the definition of α and $(2 - \mu) < s$ to arrive at

$$\int_{B_r(x_0)} \left(1 + |\nabla u_\delta|^2\right)^\alpha \, dx \leq \frac{c}{\rho^{\beta_3}} \left(1 + \int_{B_{2R}(x_0)} |\nabla u_\delta|^s \, dx\right)^{\beta_4}$$
$$+ c \int_{T_{r,\rho}} \left(1 + |\nabla u_\delta|^2\right)^\alpha \, dx$$

for some positive numbers c, β_3, β_4 which are not depending on δ. With this inequality the well known "hole filling technique" can be applied, i.e.

$$\int_{B_r(x_0)} \left(1 + |\nabla u_\delta|^2\right)^\alpha \, dx \leq \frac{\tilde{c}}{\rho^{\beta_3}} \left(1 + \int_{B_{2R}(x_0)} |\nabla u_\delta|^s \, dx\right)^{\beta_4}$$
$$+ \vartheta \int_{B_{r+\rho}(x_0)} \left(1 + |\nabla u_\delta|^2\right)^\alpha \, dx$$

holds with some $0 < \vartheta = \frac{c}{c+1} < 1$. This estimate finally proves the lemma (for $n \geq 3$) if we recall the assumptions (4) and (5) of Section 3.2 and if we refer to (see [Gia3], Lemma 5.1, p. 81)

Lemma 3.22. *Let $f(t)$ be a non-negative bounded function on $0 \leq T_0 \leq t \leq T_1$. Suppose that for $T_0 \leq t < s \leq T_1$ we have*

$$f(t) \leq A(s - t)^{-\alpha} + B + \theta f(s)$$

with $\alpha > 0$, $0 \leq \theta < 1$ and A, B non-negative constants. Then there exists a constant $c = c(\alpha, \theta)$, such that for all $\rho < R$, $T_0 \leq \rho < R \leq T_1$ we have

$$f(\rho) \leq c\left[A(R - \rho)^{-\alpha} + B\right] .$$

Now let $n = 2$ and define $\alpha = \chi(2 - \mu)/2$. Then we have

$$\int_{B_r(x_0)} \left(1 + |\nabla u_\delta|^2\right)^\alpha \, dx \leq \int_{B_{2R}(x_0)} (\eta \, h_\delta)^{2\chi} \, dx$$
$$\leq c \left(\int_{B_{2R}(x_0)} |\nabla(\eta \, h_\delta)|^t \, dx\right)^{\frac{2\chi}{t}} ,$$

where $t \in (1, 2)$ is defined through $2\chi = 2t/(2 - t)$. Using Hölder's inequality we get

$$\int_{B_r(x_0)} \left(1 + |\nabla u_\delta|^2\right)^\alpha \, dx \leq c \left(\int_{B_{2R}(x_0)} |\nabla(\eta \, h_\delta)|^2 \, dx\right)^\chi ,$$

and we can proceed as before with $n/(n - 2)$ replaced by χ. Again we have to impose the condition $(1 - \theta)q/(2 - \mu) < 1$ which for $n = 2$ is equivalent to $\chi > s/(s + 2 - \mu - q)$. But the latter inequality follows from our choice of χ, thus Lemma 3.17 is also proved in case $n = 2$. $\square$

Remark 3.23. *Once the uniform L^q-estimates are established, the weaker condition (16) is sufficient to complete the proof of both case i) and ii) as stated in Theorem 3.11.*

3.3.3 Proof of Theorem 3.16

In the following we suppose that the structural condition (11) is satisfied. Moreover, we introduce the notation

$$\omega_\delta = \ln\left(1 + |\nabla u_\delta|^2\right),$$

$$A(h,r) = A_\delta(h,r) = \{x \in B_r(x_0) : \omega_\delta > h\}, \quad h > 0,$$

where we assume from now on that $0 < r < R < R_0$. The point $x_0 \in \hat\Omega$ and the radius $R_0 > 0$ are still fixed such that $B_{2R_0}(x_0) \Subset \hat\Omega$. The first auxiliary result is given by

Lemma 3.24. *Consider $\eta \in C_0^1(B_R(x_0))$, $0 \le \eta \le 1$. Then we have for any $k > 0$*

$$\int_{A(k,R)} \left(1 + |\nabla u_\delta|^2\right)^{1-\frac{\mu}{2}} |\nabla\omega_\delta|^2 \eta^2 \, dx$$

$$+ \int_{A(k,R)} \left(1 + |\nabla u_\delta|^2\right)^{-\frac{\mu}{2}} (\omega_\delta - k)^2 \eta^2 |\nabla^2 u_\delta|^2 \, dx \tag{27}$$

$$\le c \int_{A(k,R)} \left(1 + |\nabla u_\delta|^2\right)^{\frac{q}{2}} (\omega_\delta - k)^2 |\nabla\eta|^2 \, dx.$$

Here the constant $c = c(f) < +\infty$ only depends on the data and is, in particular, independent of δ and k.

Proof of Lemma 3.24. i) Recall that in the scalar case u_δ is of class $W_{t,loc}^2(\hat\Omega)$ for any finite t. In the vectorial setting the cases $q \ge 2$ and $q < 2$ have to be distinguished. On account of (11) we may apply [GiaM], Theorem 3.1, to obtain in the case $q \ge 2$

$$\nabla u_\delta \in W_{2,loc}^1\left(\hat\Omega; \mathbb{R}^{nN}\right) \cap L_{loc}^\infty\left(\hat\Omega; \mathbb{R}^{nN}\right). \tag{28}$$

If $q < 2$, then the same result follows from [AF3], Proposition 2.7. As a consequence, replacing $\varphi \in C_0^\infty(\hat\Omega; \mathbb{R}^N)$ by $\partial_\gamma\varphi$, $\gamma = 1, \ldots, n$, the Euler equation (17) yields

$$\int_{\hat\Omega} D^2 f_\delta(\nabla u_\delta)\left(\partial_\gamma\nabla u_\delta, \nabla\varphi\right) dx = 0 \tag{29}$$

for all $\varphi \in C_0^\infty(\hat\Omega; \mathbb{R}^N)$ and for any $\gamma = 1, \ldots, n$. Moreover, (28) proves that $D^2 f_\delta(\nabla u_\delta)$ is of class L^∞, hence standard approximation arguments show that (29) is valid for any $\varphi \in W_2^1(\hat\Omega; \mathbb{R}^N)$ which is compactly supported in $\hat\Omega$.

Next we recall (see, for instance, [GT], Lemma 7.6, p. 152): given a weakly differentiable function $w \colon \hat\Omega \to \mathbb{R}$ and writing $w^+ = \max[w, 0]$ we have

$$
\nabla w^+ = \begin{cases} \nabla w & \text{if } w > 0 \,, \\ 0 & \text{if } w \le 0 \,. \end{cases}
$$

Altogether, the above listed properties of ∇u_δ show that $\varphi = \eta^2 \partial_\gamma u_\delta \max[\omega_\delta - k, 0]$ is an admissible in (29), hence

$$
\int_{A(k,R)} D^2 f_\delta(\nabla u_\delta) \Big(\partial_\gamma \nabla u_\delta, \nabla\{\eta^2 \partial_\gamma u_\delta(\omega_\delta - k)\}\Big)\, dx = 0, \quad \text{for any } k > 0 \,,
$$

or equivalently

$$
\begin{aligned}
&\int_{A(k,R)} D^2 f_\delta(\nabla u_\delta) \big(\partial_\gamma \nabla u_\delta, \partial_\gamma \nabla u_\delta\big) (\omega_\delta - k) \eta^2 \, dx \\
&\quad + \int_{A(k,R)} D^2 f_\delta(\nabla u_\delta) \big(\partial_\gamma \nabla u_\delta, \partial_\gamma u_\delta \otimes \nabla \omega_\delta\big) \eta^2 \, dx \qquad (30) \\
&= -2 \int_{A(k,R)} D^2 f_\delta(\nabla u_\delta) \big(\partial_\gamma \nabla u_\delta, \partial_\gamma u_\delta \otimes \nabla \eta\big) (\omega_\delta - k) \eta \, dx \,.
\end{aligned}
$$

The first integral on the left-hand side is non-negative and omitted in the following estimates. Studying the second one, we first consider the scalar case and immediately obtain

$$
\begin{aligned}
&\int_{A(k,R)} D^2 f_\delta(\nabla u_\delta) \big(\partial_\gamma \nabla u_\delta, \partial_\gamma u_\delta \nabla \omega_\delta\big) \eta^2 \, dx \\
&\quad = \frac{1}{2} \int_{A(k,R)} D^2 f_\delta(\nabla u_\delta)(\nabla \omega_\delta, \nabla \omega_\delta) \big(1 + |\nabla u_\delta|^2\big) \eta^2 \, dx \qquad (31) \\
&\quad \ge c \int_{A(k,R)} \big(1 + |\nabla u_\delta|^2\big)^{1 - \frac{\mu}{2}} |\nabla \omega_\delta|^2 \eta^2 \, dx \,.
\end{aligned}
$$

In the vectorial setting we have to use the structure condition (11). To this purpose assume that $\psi \colon \hat\Omega \to \mathbb{R}$ is a weakly differentiable function, let $f_\delta(Z) = g_\delta(|Z|^2)$ and denote by e_j, $j = 1, \ldots, N$, the jth coordinate vector. Using (12) we get almost everywhere

$$
\begin{aligned}
&D^2 f_\delta(\nabla u_\delta) \big(\partial_\gamma \nabla u_\delta, \partial_\gamma u_\delta \otimes \nabla \psi\big) \\
&\quad = 4 g_\delta'' \partial_\alpha u_\delta^i \partial_\gamma \partial_\alpha u_\delta^i \partial_\beta u_\delta^j \partial_\gamma u_\delta^j \partial_\beta \psi + 2 g_\delta' \partial_\gamma \partial_\alpha u_\delta^i \partial_\gamma u_\delta^i \partial_\alpha \psi \\
&\quad = 2 g_\delta'' \partial_\gamma |\nabla u_\delta|^2 \partial_\beta \psi \partial_\beta u_\delta^j \partial_\gamma u_\delta^j + g_\delta' \partial_\alpha |\nabla u_\delta|^2 \partial_\alpha \psi \qquad (32) \\
&\quad = \frac{1}{2} \frac{\partial^2 f_\delta}{\partial z_\beta^j \partial z_\gamma^j}(\nabla u_\delta) \partial_\beta \psi \partial_\gamma |\nabla u_\delta|^2 \\
&\quad = \frac{1}{2} D^2 f_\delta(\nabla u_\delta) \big(e_j \otimes \nabla \psi, e_j \otimes \nabla |\nabla u_\delta|^2\big) \,.
\end{aligned}
$$

With $\psi = \omega_\delta$, $\nabla|\nabla u_\delta|^2 = (1+|\nabla u_\delta|^2)\nabla\omega_\delta$ and the ellipticity condition (6) of Section 3.2 we arrive at the counterpart of (31) in the case $N > 1$

$$\int_{A(k,R)} D^2 f_\delta(\nabla u_\delta)(\partial_\gamma\nabla u_\delta, \partial_\gamma u_\delta \otimes \nabla\omega_\delta)\,\eta^2\,dx$$

$$= \frac{1}{2}\int_{A(k,R)} D^2 f_\delta(\nabla u_\delta)(e_j \otimes \nabla\omega_\delta, e_j \otimes \nabla\omega_\delta)(1+|\nabla u_\delta|^2)\,\eta^2\,dx \quad (33)$$

$$\geq c \int_{A(k,R)} (1+|\nabla u_\delta|^2)^{1-\frac{\mu}{2}}|\nabla\omega_\delta|^2\,\eta^2\,dx\,.$$

It remains to estimate (using (32) and Young's inequality for $\varepsilon > 0$)

$$\left|\int_{A(k,R)} D^2 f_\delta(\nabla u_\delta)(\partial_\gamma\nabla u_\delta, \partial_\gamma u_\delta \otimes \nabla\eta^2)(\omega_\delta - k)\,dx\right|$$

$$= \left|\int_{A(k,R)} D^2 f_\delta(\nabla u_\delta)(e_j \otimes \nabla\eta, e_j \otimes \nabla\omega_\delta)(1+|\nabla u_\delta|^2)(\omega_\delta - k)\eta\,dx\right|$$

$$\leq \varepsilon \int_{A(k,R)} D^2 f_\delta(\nabla u_\delta)(e_j \otimes \nabla\omega_\delta, e_j \otimes \nabla\omega_\delta)(1+|\nabla u_\delta|^2)\,\eta^2\,dx$$

$$+\frac{1}{\varepsilon}\int_{A(k,R)} D^2 f_\delta(\nabla u_\delta)(e_j \otimes \nabla\eta, e_j \otimes \nabla\eta)(\omega_\delta - k)^2(1+|\nabla u_\delta|^2)\,dx\,.$$

On account of the first equality of (33) we can absorb the ε-term if ε is sufficiently small, hence (30), (33) and (6) of Section 3.2 give

$$\int_{A(k,R)} (1+|\nabla u_\delta|^2)^{1-\frac{\mu}{2}}|\nabla\omega_\delta|^2\,\eta^2\,dx$$

$$\leq c \int_{A(k,R)} D^2 f_\delta(\nabla u_\delta)(e_j \otimes \nabla\eta, e_j \otimes \nabla\eta)(\omega_\delta - k)^2(1+|\nabla u_\delta|^2)\,dx \quad (34)$$

$$\leq c \int_{A(k,R)} (1+|\nabla u_\delta|^2)^{\frac{q}{2}}(\omega_\delta - k)^2|\nabla\eta|^2\,dx\,.$$

Inequality (34) shows that the first integral on the left-hand side of (27) is bounded as claimed in the lemma.

ii) This time we choose $\varphi = \eta^2\,\partial_\gamma u_\delta\,\big[\max[\omega - k, 0]\big]^2$ which, as above, is seen to be admissible in (29). As in (i) we get for $k > 0$

$$\int_{A(k,R)} D^2 f_\delta(\nabla u_\delta)(\partial_\gamma\nabla u_\delta, \partial_\gamma\nabla u_\delta)(\omega_\delta - k)^2\eta^2\,dx$$

$$+2\int_{A(k,R)} D^2 f_\delta(\nabla u_\delta)(\partial_\gamma\nabla u_\delta, \partial_\gamma u_\delta \otimes \nabla\omega_\delta)(\omega_\delta - k)\eta^2\,dx \quad (35)$$

$$= -2\int_{A(k,R)} D^2 f_\delta(\nabla u_\delta)(\partial_\gamma\nabla u_\delta, \partial_\gamma u_\delta \otimes \nabla\eta)(\omega_\delta - k)^2\eta\,dx\,.$$

Now the second integral on the left-hand side is non-negative which in the scalar case is immediate. If $N > 1$, then we use (32) once more to establish this observation. The right-hand side of (35) is bounded via

$$
|\text{r.h.s.}| \leq c \left\{ \varepsilon \int_{A(k,R)} D^2 f_\delta(\nabla u_\delta)(\partial_\gamma \nabla u_\delta, \partial_\gamma \nabla u_\delta)(\omega_\delta - k)^2 \eta^2 \, dx \right.
$$
$$
\left. + \frac{1}{\varepsilon} \int_{A(k,R)} D^2 f_\delta(\nabla u_\delta)(\partial_\gamma u_\delta \otimes \nabla \eta, \partial_\gamma u_\delta \otimes \nabla \eta)(\omega_\delta - k)^2 \, dx \right\} ,
$$

and by choosing $\varepsilon > 0$ properly we get from (35)

$$
\int_{A(k,R)} D^2 f_\delta(\nabla u_\delta)(\partial_\gamma \nabla u_\delta, \partial_\gamma \nabla u_\delta)(\omega_\delta - k)^2 \eta^2 \, dx
$$
$$
\leq c \int_{A(k,R)} D^2 f_\delta(\nabla u_\delta)(\partial_\gamma u_\delta \otimes \nabla \eta, \partial_\gamma u_\delta \otimes \nabla \eta)(\omega_\delta - k)^2 \, dx .
$$

This completes the proof of Lemma 3.24 if we recall the assumption (6), Section 3.2. $\square$

Finally we introduce the notation

$$
a(h,r) = a_\delta(h,r) = \int_{A(h,r)} \left(1 + |\nabla u_\delta|^2\right)^{\frac{q}{2}} \, dx ,
$$
$$
\tau(h,r) = \tau_\delta(h,r) = \int_{A(h,r)} \left(1 + |\nabla u_\delta|^2\right)^{\frac{q}{2}} (\omega_\delta - h)^2 \, dx ,
$$

to obtain

Lemma 3.25. *Suppose that $\chi > 1$ is given according to Lemma 3.17. If $h > k > 0$, then we have for a suitable constant $c = c(f)$*

$$
i) \qquad \tau(h,r) \leq c a(h,r)^{\frac{\chi-1}{\chi}} (R-r)^{-2} \tau(h,R) ;
$$

$$
ii) \qquad a(h,r) \leq (h-k)^{-2} \tau(k,r) .
$$

Proof of Lemma 3.25. The assertion *ii)* is immediate by the observation (recall the definition of $A(h,r)$)

$$
\int_{A(h,r)} \left(1 + |\nabla u_\delta|^2\right)^{\frac{q}{2}} \, dx \leq \int_{A(h,r)} \left(1 + |\nabla u_\delta|^2\right)^{\frac{q}{2}} \frac{(\omega_\delta - k)^2}{(h-k)^2} \, dx .
$$

ad i). We consider $\eta \in C_0^1(B_R(x_0))$ such that $\eta \equiv 1$ on $B_r(x_0)$, $0 \leq \eta \leq 1$, $|\nabla \eta| \leq c (R-r)^{-1}$. Moreover, let $\Gamma_\delta = \Gamma_\delta(\nabla u_\delta) = 1 + |\nabla u_\delta|^2$ and choose $\beta \in [0, \frac{q}{2})$ to be fixed later. Then we have by Sobolev's inequality in the case $n \geq 3$ (recall that $\nabla(\omega_\delta - h)^+ = 0$ on $[(\omega_\delta - h) \leq 0]$)

$$\int_{A(h,r)} \Gamma_\delta^{\frac{q}{2}} (\omega_\delta - h)^2 \, dx = \int_{A(h,r)} \Gamma_\delta^{\frac{q}{2}-\beta} (\omega_\delta - h)^2 \Gamma_\delta^\beta \, dx$$

$$\leq \left(\int_{A(h,r)} \Gamma_\delta^{(\frac{q}{2}-\beta)\chi} (\omega_\delta - h)^{2\chi} \, dx \right)^{\frac{1}{\chi}}$$

$$\cdot \underbrace{\left(\int_{A(h,r)} \Gamma_\delta^{\frac{\chi}{\chi-1}\beta} \, dx \right)^{\frac{\chi-1}{\chi}}}_{=:X_\delta}$$

$$\leq X_\delta \left(\int_{A(h,R)} \left\{ \eta \Gamma_\delta^{\frac{1}{2}(\frac{q}{2}-\beta)} (\omega_\delta - h) \right\}^{2\chi} \, dx \right)^{\frac{1}{\chi}}$$

$$\leq c X_\delta \int_{A(h,R)} \left| \nabla \left\{ \eta (\omega_\delta - h) \Gamma_\delta^{\frac{1}{2}(\frac{q}{2}-\beta)} \right\} \right|^2 \, dx \, ,$$

and the remaining integral splits into the sum of the following terms:

$$\int_{A(h,R)} |\nabla \eta|^2 (\omega_\delta - h)^2 \Gamma_\delta^{\frac{q}{2}-\beta} \, dx \leq c(R-r)^{-2} \tau(h,R) \, ;$$

$$\int_{A(h,R)} \eta^2 |\nabla \omega_\delta|^2 \Gamma_\delta^{\frac{q}{2}-\beta} \, dx \qquad \leq \text{r.h.s. of inequality (27)} \, ,$$

$$\text{provided } \frac{q}{2} - \beta \leq 1 - \frac{\mu}{2} \, ;$$

$$\int_{A(h,R)} \eta^2 (\omega_\delta - h)^2 \Gamma_\delta^{\frac{q}{2}-\beta-2} |\nabla \Gamma_\delta|^2 \, dx$$

$$\leq c \int_{A(h,R)} \eta^2 (\omega_\delta - h)^2 \Gamma_\delta^{\frac{q}{2}-\beta-1} |\nabla^2 u_\delta|^2 \, dx \leq \text{r.h.s. of (27)},$$

if again the above inequality holds for β. So let us define $\beta = \frac{1}{2}(q + \mu) - 1$, where $q \geq 2 - \mu$ shows $\beta \geq 0$. Finally

$$X_\delta \leq a(h,r)^{\frac{\chi-1}{\chi}}$$

follows from assumption (16) – of course the stronger assumption (14) also gives the result.

The modifications in the case $n = 2$ are outlined at the end of the proof of Lemma 3.17. Here we choose χ sufficiently large such that

$$\frac{\chi}{\chi-1} \left(\frac{q+\mu}{2} - 1 \right) \leq \frac{q}{2} \, .$$

This is possible on account of $\mu < 2$. Altogether we have proved Lemma 3.25.
$\square$

Given Lemma 3.25, the well known DeGiorgi-type technique is applied – we refer to [St], Lemma 5.1, p. 219.

Lemma 3.26. *Assume that $\varphi(h, \rho)$ is a non-negative real valued function defined for $h > k_0$ and $\rho < R_0$. Suppose further that for fixed ρ the function is non-increasing in h and that it is non-decreasing in ρ if h is fixed. Then*

$$\varphi(h, \rho) \leq \frac{C}{(h-k)^\alpha (R-\rho)^\gamma} \left[\varphi(k, R)\right]^\beta , \quad h > k > k_0 , \quad \rho < R < R_0 ,$$

with some positive constants C, α, $\beta > 1$, γ, implies for all $0 < \sigma < 1$

$$\varphi(k_0 + d, R_0 - \sigma R_0) = 0 ,$$

where the quantity d is given by

$$d^\alpha = \frac{2^{(\alpha+\beta)\beta/(\beta-1)} C \left[\varphi(k_0, R_0)\right]^{\beta-1}}{\sigma^\gamma R_0^\gamma} .$$

Note that Lemma 3.25 implies: for all $h > k > 0$, $0 < r < R < R_0$ we have

$$\tau(h, r) \leq \frac{c}{(R-r)^2 (h-k)^{2\frac{\varkappa-1}{\varkappa}}} \left[\tau(k, r)\right]^{\frac{\varkappa-1}{\varkappa}} \tau(h, R)$$

$$\leq \frac{c}{(R-r)^2 (h-k)^{2\frac{\varkappa-1}{\varkappa}}} \left[\tau(k, R)\right]^{1+\frac{\varkappa-1}{\varkappa}} .$$

Hence it follows from Lemma 3.26 that

$$\tau\left(d, R_0/2\right) = \int_{A(d, R_0/2)} \left(1 + |\nabla u_\delta|^2\right)^{\frac{q}{2}} (\omega_\delta - d)^2 \, \mathrm{d}x = 0$$

and, as a consequence,

$$\omega_\delta \leq d \quad \text{on } B_{R_0/2}(x_0) . \tag{36}$$

Lemma 3.25 shows that d is depending on the following quantities

$$d = d\left(f, \tau(0, R_0), R_0\right) .$$

Recalling the definition

$$\tau(0, R_0) = \int_{B_{R_0}(x_0)} \left(1 + |\nabla u_\delta|^2\right)^{\frac{q}{2}} \omega_\delta^2 \, \mathrm{d}x$$

we may write using Lemma 3.17

$$d = d\left(f, J_\delta[u_\delta], R_0\right) . \tag{37}$$

Finally, the definition of ω_δ completes the proof of Theorem 3.16 since we have established (36) and (37). $\square$

3.3.4 Conclusion

In the last step we are going to prove Theorem 3.11 as a corollary of Theorem 3.16. Let us suppose first that we have the additional growth estimate

$$c_1 F(|Z|) - c_2 \leq f(Z) \leq c_3 \left(1 + F(|Z|)\right) \tag{38}$$

for all $Z \in \mathbb{R}^{nN}$ and for some real numbers c_1, c_2, c_3. Moreover, let us assume that

$$u_0 \in W_q^1\left(\Omega; \mathbb{R}^N\right) . \tag{39}$$

Then, at the beginning of Section 3.3.1, we may choose $\hat{u}_0 = u_0$, $\hat{\Omega} = \Omega$. Furthermore, we claim that in this case u_δ is a J-minimizing sequence in $u_0 + \overset{\circ}{W}_F^1(\Omega; \mathbb{R}^N)$. In fact, for any $w \in \overset{\circ}{W}_F^1(\Omega; \mathbb{R}^N)$ there is a sequence $\{w_m\}$ in $C_0^\infty(\Omega; \mathbb{R}^N)$ such that

$$\|w - w_m\|_{W_F^1(\Omega; \mathbb{R}^N)} \to 0 \quad \text{as } m \to \infty . \tag{40}$$

Now, if we have the assumption (39), then (18) gives a uniform bound for $J_\delta[u_\delta]$. Hence there is a function $\tilde{u} \in u_0 + \overset{\circ}{W}_F^1(\Omega; \mathbb{R}^N)$ satisfying (after passing to a subsequence) as $\delta \to 0$

$$u_\delta \to \tilde{u} \text{ in } L^1\left(\Omega; \mathbb{R}^N\right) , \quad \nabla u_\delta \rightharpoonup \nabla \tilde{u} \text{ in } L^1\left(\Omega; \mathbb{R}^{nN}\right) . \tag{41}$$

Lower semicontinuity and the minimality of u_δ give for any $\varphi \in C_0^\infty(\Omega; \mathbb{R}^N)$

$$J[\tilde{u}] \leq \liminf_{\delta \to 0} J[u_\delta] \leq \liminf_{\delta \to 0} J_\delta[u_\delta] \leq \limsup_{\delta \to 0} J_\delta[u_0 + \varphi]$$
$$= J[u_0 + \varphi] . \tag{42}$$

We finally note that the functional J is convex and locally bounded from above on $W_F^1(\Omega; \mathbb{R}^N)$. This is a consequence of the right-hand inequality of (38). Hence J is continuous on $W_F^1(\Omega; \mathbb{R}^N)$ and the density result (40) together with (42) proves $\tilde{u} = u$, where u is the unique solution of the variational problem $(\mathcal{P})$. Theorem 3.16 and (41) imply

$$\|\nabla u\|_{L_{loc}^\infty(\Omega; \mathbb{R}^{nN})} \leq const. \tag{43}$$

Up to now (43) merely is proved if we have the additional assumptions (38) and (39). Without these assumptions we do not obtain a uniform bound for $J_\delta[u_\delta]$ (recall (18)). Moreover, we cannot apply the density result (40) any longer since we lose the continuity of J on $W_F^1(\Omega; \mathbb{R}^N)$.

To overcome these difficulties we consider an ε-mollification $(u)_\varepsilon$ of u and choose $\hat{u}_0 = (u)_\varepsilon$ and $\hat{\Omega} = B$ for some ball $B \Subset \Omega$. Here $\varepsilon > 0$ is supposed to be sufficiently small. Now we have $u_\delta = u_\delta^\varepsilon$. If $\delta = \delta(\varepsilon)$ is given by

$$\delta = \delta(\varepsilon) = \frac{1}{1 + \varepsilon^{-1} + \|\nabla(u)_\varepsilon\|_{L^q(B, \mathbb{R}^{nN})}^{2q}} \, ,$$

then the minimality of u_δ implies

$$\int_B f(\nabla u_\delta) \, dx \leq \int_B f_\delta(\nabla u_\delta) \, dx \leq \int_B f_\delta(\nabla(u)_\varepsilon) \, dx$$

and from Jensen's inequality we deduce

$$\int_B f(\nabla(u)_\varepsilon) \, dx \leq \int_B f(\nabla u) \, dx + O(\varepsilon) \, ,$$

where $O(\varepsilon) \to 0$ as $\varepsilon \to 0$. Next we claim

$$\delta(\varepsilon) \int_B (1 + |\nabla(u)_\varepsilon|^2)^{\frac{q}{2}} \, dx \leq c(R) \sqrt{\varepsilon} \tag{44}$$

with $c(R)$ independent of ε. For the proof we observe that by the definition of $\delta(\varepsilon)$ the left-hand side of (44) is dominated by $c(R) \frac{1+x}{1+x^2+\varepsilon^{-1}}$, $x := \int_B |\nabla(u)_\varepsilon|^q dx$.

Case 1: If $x \leq \frac{1}{\sqrt{\varepsilon}}$, then

$$\frac{1+x}{1+x^2+\varepsilon^{-1}} \leq \frac{1+\sqrt{\varepsilon}^{-1}}{1+x^2+\varepsilon^{-1}} \leq \frac{1+\sqrt{\varepsilon}^{-1}}{1+\varepsilon^{-1}} = \frac{\varepsilon+\sqrt{\varepsilon}}{\varepsilon+1}$$

$$\leq \varepsilon + \sqrt{\varepsilon} \leq 2\sqrt{\varepsilon} \, .$$

Case 2: Consider the case $x \geq \frac{1}{\sqrt{\varepsilon}}$. Using $1 + x^2 \geq \frac{1}{2}(1+x)^2$ we obtain

$$\frac{1+x}{1+x^2+\varepsilon^{-1}} \leq \frac{1+x}{1+x^2} \leq \frac{2}{1+x} \leq \frac{2}{x} \leq 2\sqrt{\varepsilon} \, ,$$

and (44) is established.

Thus, on one hand we have by (18) a uniform bound for $J_\delta[u_\delta]$. On the other hand it is proved that

$$\int_B F\big(|\nabla u_\delta|\big) \, dx \leq \int_B f(\nabla u_\delta) \, dx \leq \int_B f(\nabla u) \, dx + O(\varepsilon) \, . \tag{45}$$

Hence, on B there exists a weak W_1^1-limit $\tilde{u}$ of the sequence u_δ. Lower semicontinuity and (45) give $\tilde{u} = u$ on B. As already mentioned above, $J_\delta[u_\delta]$ is uniformly bounded (with our choice of $\delta(\varepsilon)$) and Theorem 3.16 implies the claim (43) in the general setting as well.

Hölder continuity of the gradient then follows by well known arguments from the Euler equation

$$\int_\Omega \nabla f(\nabla u) : \nabla \varphi \, dx = 0 \quad \text{for all } \varphi \in C_0^1(\Omega; \mathbb{R}^N) \, .$$

In the scalar case $N = 1$ we argue with the standard difference quotient technique and, since u is Lipschitz, it follows that u is of class $W^2_{2,loc}(\Omega)$. Thus, letting $v = \partial_\gamma u$, one arrives at

$$\int_\Omega D^2 f(\nabla u)(\nabla v, \nabla \varphi) \, dx = 0 \quad \text{for all } \varphi \in C^1_0(\Omega)$$

(compare Section 3.3.1), where the coefficients $\frac{\partial^2 f}{\partial z_\alpha \partial z_\beta}(\nabla u)$ are uniformly elliptic on $\Omega' \Subset \Omega$. Theorem 8.22, p. 200, of [GT] gives the Hölder continuity of v.

In the vector-valued case an auxiliary integrand $\tilde{f}$ is constructed following the lines of [MS]. As a result, Theorem 3.1 of [GiaM] can be applied since we also have supposed the Hölder condition (13). Thus, the $C^{1,\alpha}$-regularity is proved in the vectorial setting as well. $\square$

3.4 Partial regularity in the general vectorial setting

The notion of (s, μ, q)-growth was introduced as an appropriate assumption to obtain local $C^{1,\alpha}$-regularity of the unique minimizer of the variational problem $(\mathcal{P})$ in the scalar case. Moreover, this result holds true in the vectorial setting with some additional structure. Going through the proof of Theorem 3.11 we already have recognized that uniform L^q_{loc}-estimates in the sense of Lemma 3.17 do not depend on this special structure. Thus we expect that the notion of (s, μ, q)-growth is a suitable one to provide partial regularity in the general vectorial setting. In fact, we have

Theorem 3.27. *Let f be an integrand of (s, μ, q)-growth satisfying (14) or (15) & (16), respectively, as stated in Theorem 3.11. If u denotes the unique minimizer of the problem $(\mathcal{P})$ with $u_0 \in \mathbb{J}_F$, then there is an open subset $\Omega_0 \subset \Omega$ of full measure, i.e. $|\Omega - \Omega_0| = 0$, such that $u \in C^{1,\alpha}(\Omega_0, \mathbb{R}^N)$ for any $0 < \alpha < 1$. Here we do not assume the structure conditions (11) and (13) from Section 3.3.*

The *Proof of Theorem 3.27* splits into four parts, the regularization, a Caccioppoli-type inequality, a blow-up lemma and a standard iteration argument.

3.4.1 Regularization

Given $\varepsilon > 0$ we define $(u)_\varepsilon$ as the ε-mollification of u through a family of smooth mollifiers. Moreover, we fix $x_0 \in \Omega$ and $R > 0$ such that $B_R(x_0) \subset \{x \in \Omega : \text{dist}(x, \partial\Omega) > \varepsilon\}$. Then we define the regularization $u_\delta = u^\varepsilon_\delta$ as in Section 3.3.1 where we now take $\hat{u}_0 = (u)_\varepsilon$ and $\hat{\Omega} = B_R(x_0)$. A comparison with the second case discussed in the Conclusion 3.3.4 yields

Lemma 3.28. *If $\delta = \delta(\varepsilon)$ is chosen sufficiently small and if we again write $u_\delta = u_\delta^\varepsilon$, $f_\delta = f_{\delta(\varepsilon)}$, then we have as $\delta \to 0$*

(in order to use the same symbols as in the sections, where no additional parameter ε is needed, we write in the following – with a slight abuse of notation – "$\delta \to 0$"; to be more precise we should write: "$\varepsilon \to 0$ and $\delta(\varepsilon)$ is chosen sufficiently small")

$$i) \qquad\qquad\qquad u_\delta \to u \quad in \ W_1^1\big(B_R(x_0), \mathbb{R}^N\big) \,,$$

$$ii) \quad \delta \int_{B_R(x_0)} \big(1 + |\nabla u_\delta|^2\big)^{\frac{q}{2}} \, dx \to 0 \,,$$

$$iii) \qquad \int_{B_R(x_0)} f(\nabla u_\delta) \, dx \to \int_{B_R(x_0)} f(\nabla u) \, dx \,,$$

$$iv) \qquad \int_{B_R(x_0)} f_\delta(\nabla u_\delta) \, dx \to \int_{B_R(x_0)} f(\nabla u) \, dx \,.$$

3.4.2 A Caccioppoli-type inequality

In Lemma 3.17 we already have established higher integrability of ∇u. Now we proceed with an auxiliary proposition in order to obtain the limit-version of the Caccioppoli-type inequality Lemma 3.19.

Proposition 3.29. *Let $\Theta(t) = (1 + t^2)^{\frac{2-\mu}{4}}$, $t \geq 0$, $h_\delta = \Theta(|\nabla u_\delta|)$, $h = \Theta(|\nabla u|)$. Then*

$$i) \quad h \in W_{2,loc}^1\big(B_R(x_0)\big) \,,$$

$$ii) \quad h_\delta \to h \quad in \ W_{2,loc}^1\big(B_R(x_0)\big) \ as \ \delta \to 0 \,,$$

$$iii) \quad \nabla u_\delta \to \nabla u \quad almost \ everywhere \ on \ B_R(x_0) \ as \ \delta \to 0 \,.$$

Proof. We fix $0 < r < \hat{r} < R$, then we combine Lemma 3.19 and Proposition 3.20 to obtain

$$\|\nabla h_\delta\|_{L^2(B_r(x_0), \mathbb{R}^n)}^2 \leq c\big(1 + \|\nabla u_\delta\|_{L^q(B_{\hat{r}}(x_0), \mathbb{R}^{nN})}^q\big) \,.$$

Hence, by Lemma 3.17, h_δ is uniformly bounded in $W_{2,loc}^1(B_R(x_0))$ and we may assume as $\delta \to 0$

$$h_\delta \to: \hat{h} \quad \text{weakly in } W_{2,loc}^1\big(B_R(x_0)\big) \text{ and almost everywhere.}$$

To prove the pointwise convergence stated in *iii)*, we follow [FO], Lemma 4.1, and write

$$\int_{B_R(x_0)} \big(f(\nabla u_\delta) - f(\nabla u)\big)\, dx$$

$$= \int_{B_R(x_0)} \nabla f(\nabla u) : (\nabla u_\delta - \nabla u)\, dx \tag{46}$$

$$+ \int_{B_R(x_0)} \int_0^1 D^2 f\big(\nabla u + t(\nabla u_\delta - \nabla u)\big)$$

$$(\nabla u_\delta - \nabla u, \nabla u_\delta - \nabla u)\,(1-t)\, dt\, dx\ .$$

Note that on account of $u \in W^1_{q,loc}(\Omega; \mathbb{R}^N)$ and $u_\delta \in W^1_q(B_R(x_0); \mathbb{R}^N)$ the quantities on the right-hand side of (46) are well defined. Recall that $u_\delta \in (u)_\varepsilon + \overset{\circ}{W}^1_q(B_R(x_0); \mathbb{R}^N)$, where $(u)_\varepsilon$ was a regularization of the function u, in particular we have

$$\|u - (u)_\varepsilon\|_{W^1_q(\tilde\Omega; \mathbb{R}^N)} \overset{\varepsilon \to 0}{\to} 0 \quad \text{for all } \tilde\Omega \Subset \Omega\ . \tag{47}$$

Moreover,

$$\int_{B_R(x_0)} \nabla f(\nabla u) : (\nabla u_\delta - \nabla u)\, dx$$

$$= \int_{B_R(x_0)} \nabla f(\nabla u) : (\nabla u_\delta - \nabla (u)_\varepsilon)\, dx$$

$$+ \int_{B_R(x_0)} \nabla f(\nabla u) : (\nabla (u)_\varepsilon - \nabla u)\, dx\ ;$$

the first term on the right-hand side is zero due to the Euler equation satisfied by u, (47) shows that second one vanishes as $\delta \to 0$ (which means $\varepsilon \to 0$ and $\delta(\varepsilon)$ is chosen sufficiently small, compare Lemma 3.28), thus

$$\int_{B_R(x_0)} \nabla f(\nabla u) : (\nabla u_\delta - \nabla u)\, dx \to 0 \quad \text{as } \delta \to 0\ . \tag{48}$$

In addition, by Lemma 3.28, $iii)$, the left-hand side of (46) converges to zero and it follows from (46) that

$$\int_{B_R(x_0)} \int_0^1 D^2 f\big(\nabla u + t(\nabla u_\delta - \nabla u)\big)(\nabla u_\delta - \nabla u, \nabla u_\delta - \nabla u)\,(1-t)\, dt\, dx \overset{\delta \to 0}{\to} 0\ .$$

In the case $\mu \le 0$ we immediately obtain the claim, thus let us consider the case $\mu > 0$. Then we estimate

$$\int_0^1 D^2 f\big(\nabla u + t(\nabla u_\delta - \nabla u)\big)(\nabla u_\delta - \nabla u, \nabla u_\delta - \nabla u)\,(1-t)\, dt$$

$$\ge \int_0^1 \big(1 + |\nabla u + t(\nabla u_\delta - \nabla u)|^2\big)^{-\frac{\mu}{2}} |\nabla u_\delta - \nabla u|^2\,(1-t)\, dt$$

$$\ge c\big(1 + [|\nabla u| + |\nabla u_\delta|]^2\big)^{-\frac{\mu}{2}} |\nabla u_\delta - \nabla u|^2\ .$$

Since $|\nabla u_\delta|$ admits pointwise almost everywhere a finite limit (we have almost everywhere convergence of h_δ after passing to a subsequence), the above inequality proves that $\nabla u_\delta \to \nabla u$ almost everywhere, thus iii). As a consequence, we also have $\hat{h} = h$ and the proof of Proposition 3.29 is complete. $\square$

Now we formulate the limit version of Lemma 3.19.

Lemma 3.30. *There is a real number c such that for all $\eta \in C_0^1(B_R(x_0))$, $0 \leq \eta \leq 1$, and for all $Q \in \mathbb{R}^{nN}$*

$$\int_{B_R(x_0)} \eta^2 |\nabla h|^2 \, dx \leq c\|\nabla \eta\|_\infty^2 \int_{\mathrm{spt}\,\nabla\eta} |D^2 f(\nabla u)||\nabla u - Q|^2 \, dx \,.$$

Proof. Given Q, η as above, Proposition 3.29, lower semicontinuity, Lemma 3.19 and Proposition 3.20 together imply

$$\int_{B_R(x_0)} \eta^2 |\nabla h|^2 \, dx \leq \liminf_{\delta \to 0} \int_{B_R(x_0)} \eta^2 |\nabla h_\delta|^2 \, dx$$

$$\leq \liminf_{\delta \to 0} c\|\nabla \eta\|_\infty^2 \int_{\mathrm{spt}\,\nabla\eta} |D^2 f_\delta(\nabla u_\delta)||\nabla u_\delta - Q|^2 \, dx \qquad (49)$$

$$= \liminf_{\delta \to 0} c\|\nabla \eta\|_\infty^2 \int_{\mathrm{spt}\,\nabla\eta} |D^2 f(\nabla u_\delta)||\nabla u_\delta - Q|^2 \, dx \,.$$

Here, for the last equality, we made use of Lemma 3.28, ii). Next, by the pointwise convergence almost everywhere stated in Proposition 3.29, iii), we have

$$|D^2 f(\nabla u_\delta)||\nabla u_\delta - Q|^2 \to |D^2 f(\nabla u)||\nabla u - Q|^2 \quad \text{a.e. as } \delta \to 0 \,. \qquad (50)$$

Finally, by Lemma 3.17, we know that $|D^2 f(\nabla u_\delta)|\,|\nabla u_\delta - Q|^2$ is uniformly bounded in $L_{loc}^{1+\tau}(B_R(x_0))$ for some $\tau > 0$, hence

$$|D^2 f(\nabla u_\delta)||\nabla u_\delta - Q|^2 \rightharpoonup: \vartheta \quad \text{in } L_{loc}^{1+\tau}(B_R(x_0)) \,,$$

$$\int_{B_R(x_0)} |D^2 f(\nabla u_\delta)||\nabla u_\delta - Q|^2 \, dx \to \int_{B_R(x_0)} \vartheta \, dx \qquad (51)$$

as $\delta \to 0$. From (50), (51) we clearly get $\vartheta = |D^2 f(\nabla u)|\,|\nabla u - Q|^2$, which together with (49) completes the proof of Lemma 3.30. $\square$

3.4.3 Blow-up

The next step is to prove the main decay estimate given in Lemma 3.32. Partial regularity then will follow by a standard iteration argument which is sketched in Section 3.4.4 for the readers convenience.

Depending on the cases $q \geq 2$ and $q < 2$ an appropriate excess function has to be introduced: in the case $q \geq 2$ we let for balls $B_r(x) \Subset B_R(x_0) \Subset \Omega$

$$E^+(x,r) := \fint\limits_{B_r(x)} |\nabla u - (\nabla u)_{x,r}|^2 \, dy + \fint\limits_{B_r(x)} |\nabla u - (\nabla u)_{x,r}|^q \, dy \,,$$

where $(g)_{x,r}$ denotes the mean value of the function g with respect to the ball $B_r(x)$. In the case $q < 2$ we define for all $\xi \in \mathbb{R}^k$, $k \in \mathbb{N}$,

$$V(\xi) := \left(1 + |\xi|^2\right)^{\frac{q-2}{4}} \xi \,.$$

The properties of V are studied for example in [CFM], in particular we refer to Lemma 2.1 of [CFM]:

Lemma 3.31. *Let $1 < q < 2$ and let V be defined as above. Then for any ξ, $\eta \in \mathbb{R}^k$ and for any $t > 0$ we have*

i) $\quad 2^{\frac{q-2}{4}} \min\left\{|\xi|, |\xi|^{\frac{q}{2}}\right\} \leq |V(\xi)| \leq \min\left\{|\xi|, |\xi|^{\frac{q}{2}}\right\};$

ii) $\quad |V(t\xi)| \leq \max\left\{t, t^{\frac{q}{2}}\right\} |V(\xi)|;$

iii) $\quad |V(\xi + \eta)| \leq c(q) \left[|V(\xi)| + |V(\eta)|\right];$

iv) $\quad \dfrac{q}{2}|\xi - \eta| \leq \dfrac{|V(\xi) - V(\eta)|}{(1 + |\xi|^2 + |\eta|^2)^{\frac{q-2}{4}}} \leq c(k,q)\,|\xi - \eta|;$

v) $\quad |V(\xi) - V(\eta)| \leq c(k,q)\,|V(\xi - \eta)|;$

vi) $\quad |V(\xi - \eta)| \leq c(q,M)\,|V(\xi) - V(\eta)| \quad \text{if} \ \ |\eta| \leq M \ \text{and} \ \ \xi \in \mathbb{R}^k.$

With these preliminaries we let for $q < 2$

$$E^-(x,r) := \fint\limits_{B_r(x)} |V(\nabla u(x)) - V((\nabla u)_{x,r})|^2 \, dy \,.$$

This definition makes sense since V is of growth rate $q/2$ and since we have established Lemma 3.17. In both cases $q \geq 2$ and $q < 2$ we have

Lemma 3.32. *Fix $L > 0$. Then there exists a constant $C_*(L)$ such that for every $0 < \tau < 1/4$ there is an $\varepsilon = \varepsilon(L, \tau)$ satisfying: if $B_r(x) \Subset B_R(x_0)$ and if we have*

$$|(\nabla u)_{x,r}| \leq L \,, \qquad E(x,r) \leq \varepsilon(L,\tau) \,,$$

then

$$E(x, \tau r) \leq C_*(L)\tau^2 \, E(x,r) \,.$$

Here and in the following E denotes – depending on q – E^+ or E^-, respectively.

Proof. The proof is organized in four steps, always distinguishing the cases $q \geq 2$ and $q < 2$. If $q \geq 2$ then we mostly refer to [FO], the case $q < 2$ follows the lines of [CFM] and [EM2].

3.4.3.1 Blow-up and limit equation

To argue by contradiction, assume that $L > 0$ is fixed, the corresponding constant $C_*(L)$ will be chosen later on (see Section 3.4.3.4). If Lemma 3.32 is not true, then for some $0 < \tau < 1/4$ there are balls $B_{r_m}(x_m) \Subset B_R(x_0)$ such that

$$\left|(\nabla u)_{x_m, r_m}\right| \leq L , \quad E(x_m, r_m) =: \lambda_m^2 \overset{m \to \infty}{\longrightarrow} 0 , \tag{52}$$

$$E(x_m, \tau r_m) > C_* \tau^2 \lambda_m^2 . \tag{53}$$

Now, a sequence of rescaled functions is introduced by letting

$$a_m := (u)_{x_m, r_m} , \quad A_m := (\nabla u)_{x_m, r_m} ,$$

$$u_m(z) := \frac{1}{\lambda_m r_m} \left[u(x_m + r_m z) - a_m - r_m A_m z \right] \quad \text{if } |z| \leq 1 .$$

Passing to a subsequence, which is not relabeled, (52) implies

$$A_m \to : A \quad \text{in } \mathbb{R}^{nN} . \tag{54}$$

We also observe that

$$\nabla u_m(z) = \lambda_m^{-1} \left[\nabla u(x_m + r_m z) - A_m \right] , \quad (u_m)_{0,1} = 0 , \quad (\nabla u_m)_{0,1} = 0 ,$$

and concentrate for the moment on the case $q \geq 2$. Using (52), (53) and letting $B_r := B_r(0)$ we have

$$\fint_{B_1} |\nabla u_m|^2 \, dz + \lambda_m^{q-2} \fint_{B_1} |\nabla u_m|^q \, dz = \lambda_m^{-2} E^+(x_m, r_m) = 1 , \tag{55}$$

$$\fint_{B_\tau} |\nabla u_m - (\nabla u_m)_{0,\tau}|^2 \, dz + \lambda_m^{q-2} \fint_{B_\tau} |\nabla u_m - (\nabla u_m)_{0,\tau}|^q \, dz > C_* \tau^2 . \tag{56}$$

With (55) we obtain as $m \to \infty$

$$u_m \to : \hat{u} \quad \text{in } W_2^1(B_1; \mathbb{R}^N) , \tag{57}$$

$$\lambda_m \nabla u_m \to 0 \quad \text{in } L^2(B_1; \mathbb{R}^{nN}) \text{ and almost everywhere} , \tag{58}$$

$$\lambda_m^{1-\frac{2}{q}} \nabla u_m \to 0 \quad \text{in } L^q(B_1; \mathbb{R}^{nN}) \text{ if } q > 2 . \tag{59}$$

Considering the case $q < 2$ we follow [CFM], Proposition 3.4, Step 1: Lemma 3.31, *ii*), *vi*), and (54) yield

$$\fint_{B_1} \left|V(\nabla u_m(z))\right|^2 \, dz = \fint_{B_{r_m}(x_m)} \left|V\left(\frac{\nabla u(x) - A_m}{\lambda_m}\right)\right|^2 \, dx$$

$$\leq \frac{1}{\lambda_m^2} \fint_{B_{r_m}(x_m)} \left|V(\nabla u(x) - A_m)\right|^2 \, dx \tag{60}$$

$$\leq \frac{c^2(q,L)}{\lambda_m^2} \fint_{B_{r_m}(x_m)} \left|V(\nabla u(x)) - V(A_m)\right|^2 \, dx$$

$$\leq c(L) \, .$$

Hence, the "$q/2$-growth" of V (compare Lemma 3.31, i)) implies the existence of a finite constant, independent of m, such that

$$\|\nabla u_m\|_{L^q(B_1;\mathbb{R}^{nN})} \leq c \, .$$

Thus, in the subquadratic situation (57)–(59) have to be replaced by

$$u_m \rightharpoonup: \hat{u} \quad \text{in} \quad W_q^1(B_1;\mathbb{R}^N) \, , \tag{61}$$

$$\lambda_m \nabla u_m \to 0 \quad \text{in} \quad L^q(B_1;\mathbb{R}^{nN}) \quad \text{and almost everywhere} \, . \tag{62}$$

In both cases the limit $\hat{u}$ satisfies a blow-up equation stated in

Proposition 3.33. *There is a constant C^*, only depending on L, such that for all $\varphi \in C_0^1(B_1;\mathbb{R}^N)$*

$$\int_{B_1} D^2 f(A)(\nabla \hat{u}, \nabla \varphi) \, dz = 0 \, ,$$

$$\fint_{B_\tau} \left|\nabla \hat{u} - (\nabla \hat{u})_\tau\right|^2 \, dz \leq C^* \tau^2 \, . \tag{63}$$

Proof. The proof of the limit equation is well known (compare, for instance, [Ev], [EM2]), for the sake of completeness let us sketch the main arguments: the Euler equation for u obviously implies for any $\varphi \in C_0^1(B_1;\mathbb{R}^N)$

$$\int_{B_1} \int_0^1 D^2 f(A_m + s\lambda_m \nabla u_m)(\nabla u_m, \nabla \varphi) \, ds \, dz = 0 \, ,$$

i.e. we also have with φ fixed as above

$$\int_{B_1} D^2 f(A_m)(\nabla u_m, \nabla \varphi) \, dz$$

$$= -\int_{B_1} \left[\int_0^1 D^2 f(A_m + s\lambda_m \nabla u_m)(\nabla u_m, \nabla \varphi) \, ds \right. \tag{64}$$

$$\left. - D^2 f(A_m)(\nabla u_m, \nabla \varphi) \right] \, dz \, .$$

On account of (54) and since we have (57) and (61), respectively, the left-hand side of (64) converges to

$$\int_{B_1} D^2 f(A)(\nabla \hat{u}, \nabla \varphi) \, dz$$

as $m \to \infty$. Hence, the limit-equation is proved if the right-hand side of (64) vanishes when passing to the limit $m \to \infty$. Now, if $\delta > 0$ is fixed, then (58) and (62), respectively, prove the existence of a measurable set $S \subset B_1$ such that $|B_1 - S| \leq \delta$ and such that

$$\lambda_m \nabla u_m \overset{m \to \infty}{\rightrightarrows} 0 \quad \text{on } S .$$

As a consequence – and with an application of Hölder's inequality – it only remains to show that

$$\lim_{m \to \infty} \int_{B_1 - S} \int_0^1 D^2 f(A_m + s \lambda_m \nabla u_m)(\nabla u_m, \nabla \varphi) \, ds \, dz \leq c(\delta) \overset{\delta \to 0}{\rightrightarrows} 0 . \quad (65)$$

If $|D^2 f|$ is bounded, then (65) directly follows from Hölder's inequality. If $q > 2$, then we have (considering $[|\nabla u_m| \leq (>)\delta^{-1}]$)

$$\int_{B_1 - S} \left(1 + \lambda_m^2 |\nabla u_m|^2\right)^{\frac{q-2}{2}} |\nabla u_m| \, dz$$

$$\leq c(\delta) \|\nabla u_m\|_{L^2(B_1; \mathbb{R}^{nN})} + \delta^{1-q} \lambda_m^{q-2} |B_1 - S|$$

$$+ \delta \lambda_m^{q-2} \int_{B_1 - S} |\nabla u_m|^q \, dz ,$$

hence (65) and the limit equation holds on account of (55).

Up to now it is proved that $\hat{u}$ is a weak solution of an elliptic system with constant coefficients. Applying the standard theory of linear elliptic systems (see [Gia1], Chapter 3), $\hat{u}$ is seen to be of class $C^\infty(B_1; \mathbb{R}^N)$. We like to remark that it does not matter if $q < 2$: in this case we take [CFM], Proposition 2.10 as a reference.

Finally, the Campanato estimate (63) of Proposition 3.33 is proved in [Gia1], Theorem 2.1 (compare also Remark 2.3). $\square$

3.4.3.2 An auxiliary proposition

Proceeding in the proof of Lemma 3.32 we have to show the following proposition which will imply strong convergence in the third step.

Proposition 3.34. *Let $q \geq 2$ and $0 < \rho < 1$ or consider the case $q < 2$ together with $0 < \rho < 1/3$. Then*

$$\lim_{m \to \infty} \int_{B_\rho} \left(1 + |A_m + \lambda_m \nabla \hat{u} + \lambda_m \nabla w_m|^2\right)^{-\frac{\mu}{2}} |\nabla w_m|^2 \, dz = 0 ,$$

where we have set $w_m = u_m - \hat{u}$.

Remark 3.35. *The restriction $\rho < 1/3$ in the case $q < 2$ is needed to apply the Sobolev-Poincaré-type inequality Theorem 2.4 of [CFM].*

Proof of Proposition 3.34. Again $q \geq 2$ is the first case to consider, where the basic ideas are given for example in [EG1]. Here we argue exactly as in [FO], pp. 410, i.e. we observe that for all $\varphi \in C_0^1(B_1; \mathbb{R}^N)$, $0 \leq \varphi \leq 1$,

$$\int_{B_1} \int_0^1 \varphi D^2 f\left(A_m + \lambda_m \nabla \hat{u} + s \lambda_m \nabla w_m\right)(\nabla w_m, \nabla w_m)(1 - s)\, ds\, dz$$

$$= \lambda_m^{-2} \int_{B_1} \varphi \left\{ f(A_m + \lambda_m \nabla u_m) - f(A_m + \lambda_m \nabla \hat{u}) \right\} dz \qquad (66)$$

$$- \lambda_m^{-1} \int_{B_1} \varphi \nabla f(A_m + \lambda_m \nabla \hat{u}) : \nabla w_m\, dz \,.$$

The first integral on the right-hand side of (66) can be written as

$$\int_{B_1} f(A_m + \lambda_m \nabla u_m)\, dz$$

$$- \int_{B_1} \left\{ (1 - \varphi) f(A_m + \lambda_m \nabla u_m) + \varphi f(A_m + \lambda_m \nabla \hat{u}) \right\} dz \,,$$

where the first term can be estimated using the minimality of u, the second one using the convexity of f. As a result

$$\text{l.h.s of (66)} \leq \lambda_m^{-2} \int_{B_1} f\left(A_m + \lambda_m \nabla\left[u_m + \varphi(\hat{u} - u_m)\right]\right) dz$$

$$- \lambda_m^{-2} \int_{B_1} f\left(A_m + \lambda_m \left[(1 - \varphi)\nabla u_m + \varphi \nabla \hat{u}\right]\right) dz \qquad (67)$$

$$- \lambda_m^{-1} \int_{B_1} \varphi \nabla f(A_m + \lambda_m \nabla \hat{u}) : \nabla w_m\, dz$$

$$=: I_1 - I_2 - I_3 \,.$$

With $X_m := A_m + \lambda_m((1 - \varphi)\nabla u_m + \varphi \nabla \hat{u})$ we obtain

$$I_1 - I_2 = \lambda_m^{-2} \left(\int_{B_1} f(X_m + \lambda_m (\hat{u} - u_m) \otimes \nabla \varphi)\, dz - \int_{B_1} f(X_m)\, dz \right)$$

$$= \lambda_m^{-1} \int_{B_1} \nabla f(X_m) : ((\hat{u} - u_m) \otimes \nabla \varphi)\, dz$$

$$+ \int_{B_1} \int_0^1 D^2 f(X_m + s\lambda_m (\hat{u} - u_m) \otimes \nabla \varphi) \qquad (68)$$

$$\cdot ((\hat{u} - u_m) \otimes \nabla \varphi, (\hat{u} - u_m) \otimes \nabla \varphi)(1 - s)\, ds\, dz \,.$$

In order to derive an upper bound for the last integral, we first claim that

$$\int_{\operatorname{spt} \nabla \varphi} \lambda_m^{q-2} |u_m - \hat{u}|^q \, dz \to 0 \quad \text{as } m \to \infty. \tag{69}$$

In fact, if $q = 2$ then (69) immediately follows from (57). In the case $q > 2$ we use (59) and Poincaré's inequality to obtain $\lambda_m^{1-2/q} u_m \to 0$ in $L^q(\Omega; \mathbb{R}^N)$ as $m \to \infty$. This yields together with the local gradient bound for $\nabla \hat{u}$

$$\left(\int_{\operatorname{spt} \nabla \varphi} \lambda_m^{q-2} |u_m - \hat{u}|^q \, dz \right)^{\frac{1}{q}} \leq \|\lambda_m^{1-\frac{2}{q}} \hat{u}\|_{L^q(\operatorname{spt} \nabla \varphi; \mathbb{R}^N)} + \|\lambda_m^{1-\frac{2}{q}} u_m\|_{L^q(B_1; \mathbb{R}^N)}$$

$$\to 0 \quad \text{as } m \to \infty,$$

hence (69). With the notation $O(m) \to 0$ as $m \to \infty$ we get

$$\int_{B_1} \lambda_m^{q-2} |\nabla u_m|^{q-2} |\hat{u} - u_m|^2 |\nabla \varphi|^2 \, dz$$

$$\leq \left(\lambda_m^{q-2} \int_{B_1} |\nabla u_m|^q \, dz \right)^{\frac{q-2}{q}} \left(\lambda_m^{q-2} \int_{B_1} |\hat{u} - u_m|^q |\nabla \varphi|^q \, dz \right)^{\frac{2}{q}}$$

$$= O(m),$$

thus starting from (68) we arrive at

$$\text{l.h.s of (66)} \leq c \left\{ \int_{B_1} |\nabla \varphi|^2 |w_m|^2 \, dz + \lambda_m^{q-2} \int_{B_1} |\nabla \varphi|^q |w_m|^q \, dz \right\}$$

$$+ \lambda_m^{-1} \int_{B_1} \nabla f(X_m) : ((\hat{u} - u_m) \otimes \nabla \varphi) \, dz \tag{70}$$

$$- \lambda_m^{-1} \int_{B_1} \varphi \nabla f(A_m + \lambda_m \nabla \hat{u}) : \nabla w_m \, dz + O(m).$$

Clearly, (70) corresponds to the inequality (6.6) in [FO]. Next we are going to discuss the last two integrals in (70), i.e.

$$\int_{B_1} \nabla f(X_m) : ((\hat{u} - u_m) \otimes \nabla \varphi) \, dz - \int_{B_1} \varphi \nabla f(A_m + \lambda_m \nabla \hat{u}) : \nabla w_m \, dz$$

$$= \int_{B_1} [\nabla f(X_m) - \nabla f(A_m + \lambda_m \nabla \hat{u})] : ((\hat{u} - u_m) \otimes \nabla \varphi) \, dz$$

$$- \int_{B_1} \nabla f(A_m + \lambda_m \nabla \hat{u}) : \nabla (\varphi (u_m - \hat{u})) \, dz$$

$$= I_4 - I_5.$$

Using $0 \leq \varphi \leq 1$, using the upper bound for $D^2 f$ as well as the local gradient bound for $\hat{u}$, we obtain

$$I_4 = \lambda_m \int_{B_1} \int_0^1 D^2 f\big(A_m + \lambda_m \nabla \hat{u} + s\lambda_m (1-\varphi)(\nabla u_m - \nabla \hat{u})\big)$$

$$\big(\nabla u_m - \nabla \hat{u}, (\hat{u} - u_m) \otimes \nabla \varphi\big)(1-\varphi)\, ds\, dz$$

$$\leq c\lambda_m \int_{B_1} \int_0^1 \Big(1 + \big|A_m + \lambda_m \nabla \hat{u} + s\lambda_m (1-\varphi)(\nabla u_m - \nabla \hat{u})\big|^2\Big)^{\frac{q-2}{2}}$$

$$|\nabla w_m||\nabla \varphi||w_m|\, ds\, dz$$

$$\leq c\lambda_m \left\{ \int_{B_1} |\nabla w_m||\nabla \varphi||w_m|\, dz + \lambda_m^{q-2} \int_{B_1} |\nabla w_m|^{q-1}|\nabla \varphi||w_m|\, dz \right\}.$$

For I_5 we have the equality

$$I_5 = \int_{B_1} \big[\nabla f(A_m + \lambda_m \nabla \hat{u}) - \nabla f(A_m)\big] : \nabla(\varphi w_m)\, dz$$

$$= \lambda_m \int_{B_1} \int_0^1 D^2 f(A_m + s\lambda_m \nabla \hat{u})\big(\nabla \hat{u}, \nabla(\varphi w_m)\big)\, ds\, dz,$$

hence (70) implies

$$\text{l.h.s of (66)} \leq c \left\{ \int_{B_1} |\nabla \varphi|^2 |w_m|^2\, dz + \lambda_m^{q-2} \int_{B_1} |\nabla \varphi|^q |w_m|^q\, dz \right\}$$

$$+ c \left\{ \int_{B_1} |\nabla w_m||\nabla \varphi||w_m|\, dz \right.$$

$$\left. + \lambda_m^{q-2} \int_{B_1} |\nabla w_m|^{q-1}|\nabla \varphi||w_m|\, dz \right\} \quad (71)$$

$$- \int_{B_1} \int_0^1 D^2 f(A_m + s\lambda_m \nabla \hat{u})\big(\nabla \hat{u}, \nabla(\varphi w_m)\big)\, ds\, dz$$

$$+ O(m).$$

Obviously the first integral on the right-hand side of (71) tends to zero as $m \to \infty$. The second one is already discussed in (69). The third integral vanishes when passing to the limit on account of $u_m \to \hat{u}$ in $L^2(B_1; \mathbb{R}^N)$ as $m \to \infty$. The fourth one is handled using Hölder's inequality and once again (69). Finally, the last integral is immediately seen to converge to zero on account of (57). Summarizing the results we have shown that

$$\lim_{m \to \infty} \int_{B_1} \int_0^1 \varphi D^2 f\big(A_m + \lambda_m \nabla \hat{u} + s\lambda_m \nabla w_m\big)(\nabla w_m, \nabla w_m)(1-s)\, ds\, dz = 0$$

and Proposition 3.34 is proved in the case $q \geq 2$.

In the case $q < 2$ we now benefit from [EM2] (compare [Ev]) since the proof of higher integrability given in [CFM], Step 3, is adapted to balanced structure conditions. Thus, for $\xi \in \mathbb{R}^{nN}$ we let

$$f_m(\xi) := \frac{f(A_m + \lambda_m \xi) - f(A_m) - \lambda_m \nabla f(A_m) : \xi}{\lambda_m^2}$$

and define for $0 < \rho < 1/3$, $w \in W^1_{1,loc}(B_{1/3}, \mathbb{R}^N)$

$$I_\rho^m(w) := \int_{B_\rho} f_m(\nabla w) \, dz \, .$$

Our first claim is

$$\limsup_{m \to \infty} \left\{ I_\rho^m(u_m) - I_\rho^m(\hat{u}) \right\} \leq 0 \quad \text{for almost every } \rho \in (0, 1/3) \, . \tag{72}$$

To establish (72) we fix ρ as above, we choose $0 < s < \rho$, $\eta \in C_0^\infty(B_\rho)$, $0 \leq \eta \leq 1$, $\eta \equiv 1$ on B_s, $|\nabla \eta| \leq c/(\rho - s)$, and define $\varphi_m = (\hat{u} - u_m)\eta$. On one hand, u_m obviously is a local minimizer of I_ρ^m. On the other hand we have Lemma 3.3 of [CFM]:

Lemma 3.36. *Let $f : \mathbb{R}^k \to \mathbb{R}$ be a function of class C^2 satisfying for any $\xi \in \mathbb{R}^k$*

$$|\nabla f(\xi)| \leq K \left(1 + |\xi|^2\right)^{\frac{q-1}{2}} , \quad 1 < q < 2 \, .$$

Then, for any $M > 0$ there exists a constant c depending only on M, q, K, such that if we set for any $\lambda > 0$ and $A \in \mathbb{R}^k$ with $|A| \leq M$

$$f_{A,\lambda}(\xi) = \lambda^{-2} \left[f(A + \lambda\xi) - f(A) - \lambda \nabla f(A) : \xi \right] ,$$

then

$$|f_{A,\lambda}(\xi)| \leq c(q, K, M) \left(1 + |\lambda\xi|^2\right)^{\frac{q-2}{2}} |\xi|^2 \, .$$

Recalling Remark 3.5, $iv)$, and Lemma 3.31, Lemma 3.36 yields

$$I_\rho^m(u_m) - I_\rho^m(\hat{u}) \leq I_\rho^m(u_m + \varphi_m) - I_\rho^m(\hat{u})$$

$$= \int_{B_\rho - B_s} \left[f_m(\nabla u_m + \nabla \varphi_m) - f_m(\nabla \hat{u}) \right] dz$$

$$\leq \frac{c(q, \Lambda, L)}{\lambda_m^2} \int_{B_\rho - B_s} \left[|V(\lambda_m \nabla \hat{u})|^2 \right. \tag{73}$$

$$\left. + \left|V\left(\lambda_m(\hat{u} - u_m) \otimes \nabla\eta + \lambda_m \eta \nabla\hat{u} + \lambda_m(1 - \eta)\nabla u_m\right)\right|^2\right) \right] dz$$

$$\leq \frac{c}{\lambda_m^2} \int_{B_\rho - B_s} \left[|V(\lambda_m \nabla \hat{u})|^2 + |V(\lambda_m \nabla u_m)|^2 \right.$$

$$\left. + (\rho - s)^{-2} \left|V\left(\lambda_m(\hat{u} - u_m)\right)\right|^2 \right] dz \, .$$

Next, a family of positive, uniformly bounded Radon measures μ^m on $B_{1/3}$ is introduced by letting

$$\mu^m(S) := \int_S \frac{1}{\lambda_m^2} \left[|V(\lambda_m \nabla \hat{u})|^2 + |V(\lambda_m \nabla u_m)|^2 \right] dz \ .$$

Repeating the computations outlined in (60) and using the smoothness of $\hat{u}$ on $B_{1/3}$ we obtain

$$\mu^m(B_{1/3}) = \int_{B_{1/3}} \frac{1}{\lambda_m^2} \left[|V(\lambda_m \nabla \hat{u})|^2 + |V(\lambda_m \nabla u_m)|^2 \right] dz$$

$$\leq c \left\{ 1 + \int_{B_{1/3}} \frac{1}{\lambda_m^2} |V(\lambda_m \nabla u_m)|^2 \, dz \right\} \leq c(L) \ ,$$

hence we may assume that μ^m converges in measure to a Radon measure μ on $B_{1/3}$. Moreover, without loss of generality it is supposed in the following that

$$\mu(\partial B_r) = 0 \quad \text{for any } 0 < r < 1/3 \ , \tag{74}$$

which is true for almost any r as above. Now, the Sobolev-Poincaré type inequality stated in Theorem 2.4 of [CFM] is needed:

Theorem 3.37. *If $1 < q < 2$ then there exist $\frac{2}{q} < \alpha < 2$ and $\sigma > 0$ such that if $u \in W_q^1(B_{3R}(x_0); \mathbb{R}^N)$ then*

$$\left(\fint_{B_R(x_0)} \left| V \left(\frac{u - (u)_{x_0,R}}{R} \right) \right|^{2(1+\sigma)} dx \right)^{\frac{1}{2(1+\sigma)}} \leq c \left(\fint_{B_{3R}(x_0)} |V(\nabla u)|^\alpha \, dx \right)^{\frac{1}{\alpha}} \ ,$$

where $c \equiv c(n, q, N)$ is independent of R and u.

To proceed further we fix σ as given in Theorem 3.37 and choose a real number $0 < \theta < 1$ satisfying $\frac{1}{2} = \theta + \frac{1-\theta}{2(1+\sigma)}$. Interpolation, Lemma 3.31 and Theorem 3.37 give

$$\int_{B_\rho - B_s} |V(\lambda_m (\hat{u} - u_m))|^2 \, dz$$

$$\leq \left(\int_{B_\rho - B_s} |V(\lambda_m (\hat{u} - u_m))| \, dz \right)^{2\theta}$$

$$\cdot \left(\int_{B_\rho - B_s} |V(\lambda_m (\hat{u} - u_m))|^{2(1+\sigma)} \, dz \right)^{\frac{1-\theta}{1+\sigma}}$$

$$\leq c\lambda_m^{2\theta} \left(\int_{B_1} |\hat{u} - u_m| \, dz \right)^{2\theta} \left(\int_{B_\rho} \left[|V(\lambda_m (\hat{u} - u_m)) \right.\right.$$

$$\left.\left. - \lambda_m (\hat{u} - u_m)_{0,\rho})|^{2(1+\sigma)} + |V(\lambda_m (\hat{u} - u_m)_{0,\rho})|^{2(1+\sigma)} \right] dz \right)^{\frac{1-\theta}{1+\sigma}} \leq$$

$$\leq c\lambda_m^{2\theta} \left(\int_{B_1} |\hat{u} - u_m| \, dz \right)^{2\theta}$$

$$\cdot \left[\left(\int_{B_{3\rho}} |V(\lambda_m \nabla(\hat{u} - u_m))|^2 \, dz \right)^{1+\sigma} + \lambda_m^{2(1+\sigma)} \right]^{\frac{1-\theta}{1+\sigma}} \, ,$$

and it is proved that

$$\int_{B_\rho - B_s} |V(\lambda_m(\hat{u} - u_m))|^2 \, dz \leq c\lambda_m^2 \left(\int_{B_1} |\hat{u} - u_m| \, dz \right)^{2\theta} . \qquad (75)$$

Finally, (73) and (75) imply

$$I_\rho^m(u_m) - I_\rho^m(\hat{u}) \leq c \left[\mu^m(\overline{B}_\rho - B_s) + \frac{1}{(\rho - s)^2} \left(\int_{B_1} |\hat{u} - u_m| \, dz \right)^{2\theta} \right] ,$$

hence, by taking first the limit $m \to \infty$ and then the limit $s \uparrow \rho$, we get (72) on account of (74).

Once (72) is established for some $0 < \rho < 1/3$, the following identity is the starting point to deduce a lower bound for the left-hand side:

$$I_\rho^m(u_m) - I_\rho^m(\hat{u}) = \lambda_m^{-2} \int_{B_\rho} \left[f(A_m + \lambda_m \nabla u_m) - f(A_m + \lambda_m \nabla \hat{u}) \right.$$

$$\left. -\lambda_m \nabla f(A_m) : \nabla w_m \right] dz$$

$$= \lambda_m^{-1} \int_{B_\rho} \int_0^1 \left[\nabla f(A_m + \lambda_m \nabla \hat{u} + s\lambda_m \nabla w_m) \right.$$

$$\left. -\nabla f(A_m + \lambda_m \nabla \hat{u}) \right] : \nabla w_m \, ds \, dz$$

$$+\lambda_m^{-1} \int_{B_\rho} \left[\nabla f(A_m + \lambda_m \nabla \hat{u}) - \nabla f(A_m) \right] : \nabla w_m \, dz$$

$$=: I_1 + I_2 \, .$$

The local smoothness of $\hat{u}$ immediately implies $\lim_{m \to \infty} I_2 = 0$. Since

$$I_1 = \int_{B_\rho} \int_0^1 \int_0^1 s D^2 f(A_m + \lambda_m \nabla \hat{u} + ts\lambda_m \nabla w_m)(\nabla w_m, \nabla w_m) \, dt \, ds \, dz$$

$$\geq c \int_{B_\rho} \left(1 + |A_m + \lambda_m \nabla \hat{u} + \lambda_m \nabla w_m|^2 \right)^{-\frac{\mu}{2}} |\nabla w_m|^2 \, dz$$

and since we have (72), the proposition is proved for almost all, hence for all $\rho \in (0, 1/3)$. $\square$

3.4.3.3 Strong convergence

Case i): $q \geq 2$

Proposition 3.38. *In the case* $q \geq 2$ *we have as* $m \to \infty$

$$i) \qquad \nabla u_m \to \nabla \hat{u} \quad in \ L^2_{loc}(B_1; \mathbb{R}^{nN}) \ ; \qquad (76)$$

$$ii) \quad \lambda_m^{1-\frac{2}{q}} \nabla u_m \to 0 \quad in \ L^q_{loc}(B_1; \mathbb{R}^{nN}) \quad if \ q > 2 \ .$$

Proof. Here we have to distinguish *two subcases*: for $\mu \leq 0$ the first convergence is immediate by Proposition 3.34. Using this fact, the local smoothness of $\hat{u}$ and again Proposition 3.34, the next conclusion is

$$\int_{B_\rho} \lambda_m^{-\mu} |\nabla w_m|^{2-\mu} \, dz \stackrel{m\to\infty}{\longrightarrow} 0 \quad \text{for all } 0 < \rho < 1 \ . \qquad (77)$$

The proceed further, we introduce the auxiliary functions ψ_m (see [FO]),

$$\psi_m := \lambda_m^{-1} \left[\left(1 + |A_m + \lambda_m \nabla u_m|^2\right)^{\frac{2-\mu}{4}} - \left(1 + |A_m|^2\right)^{\frac{2-\mu}{4}} \right] . \qquad (78)$$

Then, by Lemma 3.30, (57), (59) and by (6) of Section 3.2 we can estimate $(0 < \rho < 1)$

$$\int_{B_\rho} |\nabla \psi_m|^2 \, dz \leq c(\rho) \int_{B_1} |D^2 f(A_m + \lambda_m \nabla u_m)| \, |\nabla u_m|^2 \, dz \leq c(\rho) \ .$$

If we now let $\Theta(Z) := (1 + |Z|^2)^{(2-\mu)/4}$, $Z \in \mathbb{R}^{nN}$, then

$$|\psi_m| = \lambda_m^{-1} \left| \int_0^1 \frac{d}{ds} \Theta(A_m + s\lambda_m \nabla u_m) \, ds \right|$$

$$\leq c \left| \int_0^1 \nabla u_m : \nabla \Theta(A_m + s\lambda_m \nabla u_m) \, ds \right|$$

$$\leq c \int_0^1 |\nabla u_m| \left(1 + |A_m + s\lambda_m \nabla u_m|^2\right)^{-\frac{\mu}{4}} \, ds$$

$$\leq c \left(|\nabla u_m| + \lambda_m^{-\frac{\mu}{2}} |\nabla u_m|^{1-\frac{\mu}{2}} \right) .$$

With this inequality we obtain

$$\int_{B_\rho} |\psi_m|^2 \, dz \leq c(\rho) \quad \text{for all } 0 < \rho < 1 \ . \qquad (79)$$

In fact, (79) is obvious for $\mu = 0$. If $\mu < 0$, then (79) is just a consequence of (77). Thus we have proved that

$$\sup_m \|\psi_m\|_{W_2^1(B_\rho)} \leq c(\rho) < \infty \quad \text{for all } 0 < \rho < 1 \tag{80}$$

and this will imply (76), *ii*): to this purpose we fix some real number $M \gg 1$ and let $U_m = U_m(M, \rho) := \{z \in B_\rho : \lambda_m |\nabla u_m| \leq M\}$. On one hand, local L^2-convergence and $q > 2$ prove

$$
\begin{aligned}
\int_{U_m} \lambda_m^{q-2} |\nabla u_m|^q \, dz &\leq \int_{U_m} \lambda_m^{q-2} |\nabla w_m|^q \, dz + \int_{U_m} \lambda_m^{q-2} |\nabla \hat{u}|^q \, dz \\
&\leq c \int_{U_m} \lambda_m^{q-2} \left(|\nabla u_m|^{q-2} + |\nabla \hat{u}|^{q-2} \right) |\nabla w_m|^2 \\
&\quad + \int_{U_m} \lambda_m^{q-2} |\nabla \hat{u}|^q \, dz \\
&\to 0 \quad \text{as } m \to \infty .
\end{aligned}
\tag{81}
$$

On the other hand, observe that for M sufficiently large and for $z \in B_\rho - U_m$

$$\psi_m(z) \geq c \lambda_m^{-1} \lambda_m^{\frac{2-\mu}{2}} |\nabla u_m(z)|^{\frac{2-\mu}{2}} , \quad \text{i.e.}$$

$$\lambda_m^{q-2+\frac{\mu q}{2-\mu}} \psi_m^{\frac{2q}{2-\mu}}(z) \geq c \lambda_m^{q-2} |\nabla u_m(z)|^q .$$

Since (14) and (16), Section 3.3, both guarantee that $2q/(2-\mu) < 2n/(n-2)$, since by (80) ψ_m is uniformly bounded in $L^{2n/(n-2)}$ and since $q - 2 + \mu q/(2-\mu) \geq 0$ follows from $q \geq 2 - \mu$, we can conclude

$$\int_{B_\rho - U_m} \lambda_m^{q-2} |\nabla u_m|^q \, dz \to 0 \quad \text{for all } 0 < \rho < 1 \tag{82}$$

as $m \to \infty$. Summarizing the results, (81) and (82) prove Proposition 3.38 in the case $\mu \leq 0$.

Now suppose that $\underline{\mu > 0}$. Proposition 3.34 implies in the case at hand (for any $0 < \rho < 1$)

$$\int_{B_\rho} \left(1 + |\lambda_m \nabla w_m|^2 \right)^{-\frac{\mu}{2}} |\nabla w_m|^2 \, dz \to 0 \quad \text{as } m \to \infty ,$$

which immediately gives

$$\int_{U_m} |\nabla w_m|^2 \, dz \to 0 \quad \text{as } m \to \infty . \tag{83}$$

Here U_m is defined as above for fixed M and ρ. Also as above we introduce ψ_m and observe that now $|\psi_m| \leq c|\nabla u_m|$ is obvious. Thus (80) is true in the case $\mu > 0$ as well. If M is chosen sufficiently large, then

$$|\psi_m|^{\frac{4}{2-\mu}} \lambda_m^{\frac{2\mu}{2-\mu}} \geq |\nabla u_m|^2 \quad \text{on } B_\rho - U_m ,$$

and since $4/(2-\mu) \leq 2n/(n-2) \Leftrightarrow \mu \leq 4/n$ (where the last inequality is true on account of $q \geq 2$) we get

$$\int_{B_\rho - U_m} |\nabla w_m|^2 \, dz \xrightarrow{m \to \infty} 0 \quad \text{for all } 0 < \rho < 1. \tag{84}$$

With (83) and (84) the first claim of (76) also is proved in the case $\mu > 0$. The second claim (76), ii) for $\mu > 0$, follows exactly as outlined in the case $\mu \leq 0$. Hence, the proof of the proposition is complete. $\quad\square$

Case ii): $q < 2$

Proposition 3.39. *If $q < 2$, then for any $0 < \rho < 1/3$*

$$\lim_{m \to \infty} \frac{1}{\lambda_m^2} \int_{B_\rho} \left| V(\lambda_m \nabla w_m) \right|^2 \, dz = 0 \, .$$

Proof. In the subquadratic case, the auxiliary function ψ_m introduced in (78) is handled via Lemma 3.31, vi). We have

$$\int_{B_\rho} |\nabla \psi_m|^2 \, dz \leq c \int_{B_1} \left(1 + |\lambda_m \nabla u_m|^2\right)^{\frac{q-2}{2}} |\nabla u_m|^2 \, dz$$

$$\leq \frac{c}{\lambda_m^2} \fint_{B_1} \left| V(\lambda_m \nabla u_m) \right|^2 \, dz$$

$$\leq \frac{c}{\lambda_m^2} \fint_{B(x_m, R_m)} \left| V(\nabla u - A_m) \right|^2 \, dx$$

$$\leq \frac{c(L)}{\lambda_m^2} \fint_{B(x_m, R_m)} \left| V(\nabla u) - V(A_m) \right|^2 \, dx \ \leq c$$

for any $0 < \rho < 1$. In addition we have $|\psi_m| \leq c \, |\nabla u_m|$, hence $\psi_m \in W^1_{q,loc}(B_1)$, thus $\psi_m \in L^{q_1}_{loc}(B_1)$ with $q_1 := nq/(n-q)$. Iterating this argument we again have

$$\sup_m \|\psi_m\|_{W^1_2(B_\rho)} \leq c(\rho) < \infty \quad \text{for any } 0 < \rho < 1 \, . \tag{85}$$

Assume now that $0 < \rho < 1/3$. With M and U_m as before, (83) once more is a consequence of Proposition 3.34. Using Lemma 3.31 we have

$$\frac{1}{\lambda_m^2} \int_{B_\rho} \left| V(\lambda_m \nabla w_m) \right|^2 \, dz \leq \frac{c}{\lambda_m^2} \int_{U_m} \left| V(\lambda_m \nabla w_m) \right|^2 \, dz$$

$$+ \frac{c}{\lambda_m^2} \int_{B_\rho - U_m} \left| V(\lambda_m \nabla u_m) \right|^2 \, dz$$

$$+ \frac{c}{\lambda_m^2} \int_{B_\rho - U_m} \left| V(\lambda_m \nabla \hat{u}) \right|^2 \, dz \, .$$

Then, by (83),

$$\frac{1}{\lambda_m^2} \int_{U_m} |V(\lambda_m \nabla w_m)|^2 \, dz \leq \int_{U_m} |\nabla w_m|^2 \left(1 + \lambda_m^2 |\nabla w_m|^2\right)^{\frac{q-2}{2}} dz$$

$$\leq \int_{U_m} |\nabla w_m|^2 \, dz \overset{m \to \infty}{\longrightarrow} 0 \, .$$

The second term vanishes (passing to the limit $m \to \infty$) provided that

$$\int_{B_\rho - U_m} \lambda_m^{q-2} |\nabla u_m|^q \, dz \to 0 \quad \text{as } m \to \infty \, .$$

In fact, we recall the estimates for ψ_m stated after (81) – which remain valid in the case under consideration – and with the same reasoning we obtain (82), where now we make use of the a priori bound (85).

Finally, we use the local boundedness of $\nabla \hat{u}$ to get

$$\frac{1}{\lambda_m^2} \int_{B_\rho - U_m} |V(\lambda_m \nabla \hat{u})|^2 \, dz \leq \int_{B_\rho - U_m} |\nabla \hat{u}|^2 \, dz$$

$$\leq \|\nabla \hat{u}\|_{L^\infty(B_\rho, \mathbb{R}^{nN})}^2 |B_\rho - U_m| \overset{m \to \infty}{\longrightarrow} 0$$

on account of $\lambda_m \nabla u_m \to 0$ almost everywhere on B_1 as $m \to \infty$ (see (62)). This completes the proof of Proposition 3.39. $\quad\square$

3.4.3.4 Conclusion

Proposition 3.38 and (56) yield in the case $q \geq 2$

$$\fint_{B_\tau} |\nabla \hat{u} - (\nabla \hat{u})_\tau|^2 \, dz \geq C_* \tau^2 \, ,$$

thus we have a contradiction to (63) if we choose $C_* = 2C^*$.

If $q < 2$, then we estimate according to [CFM], p. 24,

$$\lim_{m \to \infty} \frac{E^-(x_m, \tau r_m)}{\lambda_m^2} = \lim_{m \to \infty} \frac{1}{\lambda_m^2} \fint_{B_{\tau r_m}(x_m)} |V(\nabla u) - V((\nabla u)_{x_m, \tau R_m})|^2 \, dz$$

$$\leq \lim_{m \to \infty} \frac{c}{\lambda_m^2} \fint_{B_{\tau r_m}(x_m)} |V(\nabla u - (\nabla u)_{x_m, \tau R_m})|^2 \, dz$$

$$= \lim_{m \to \infty} \frac{c}{\lambda_m^2} \fint_{B_\tau} |V(\lambda_m (\nabla u_m - (\nabla u_m)_{0, \tau}))|^2 \, dz \leq$$

$$\leq \lim_{m \to \infty} \frac{c}{\lambda_m^2} \int_{B_\tau} \left\{ \left|V(\lambda_m \nabla w_m)\right|^2 + \left|V\left(\lambda_m (\nabla \hat{u} - (\nabla \hat{u})_{0,\tau})\right)\right|^2 \right.$$

$$\left. + \left|V\left(\lambda_m ((\nabla \hat{u})_{0,\tau} - (\nabla u_m)_{0,\tau})\right)\right|^2 \right\} \, dz \, ,$$

where the first integral is handled using Proposition 3.39. The last one vanishes when passing to the limit $m \to \infty$ since we may first estimate

$$\int_{B_\tau} \left|V\left(\lambda_m ((\nabla \hat{u})_{0,\tau} - (\nabla u_m)_{0,\tau})\right)\right|^2 \, dz \leq \lambda_m^2 \int_{B_\tau} \left|(\nabla \hat{u})_\tau - (\nabla u_m)_\tau\right|^2 \, dz$$

and then use (61) for the right-hand side. The second integral again is estimated by (63). Thus, choosing C_* sufficiently large we also get the contradiction in the case $q < 2$ and the main decay lemma is proved. $\square$

3.4.4 Iteration

Finally, Theorem 3.27 follows from the iteration Lemma 3.40, which is a well known consequence of the main Lemma 3.32 (see, for instance, [GiuM1], [Ev] or [FH]). We include a short proof for the sake of completeness.

Lemma 3.40. *With the assumptions of Lemma 3.32 suppose that we are given numbers $\alpha \in (0,1)$ and $\tau \in (0, 1/8)$ such that*

$$C_*(2L) \tau^{2(1-\alpha)} \leq 1 \, .$$

Then there exits a number $\eta = \eta(L, \tau) > 0$ such that for every ball $B_r(x) \Subset B_R(x_0)$ the inequalities

$$\left|(\nabla u)_{x,r}\right| < L \, , \tag{86}$$
$$E(x, r) < \eta(L, \tau) \tag{87}$$

imply for any $k = 0, 1, 2, \dots$

$$\left|(\nabla u)_{x,\tau^k r}\right| \leq 2L \, , \tag{88}$$
$$E(x, \tau^k r) \leq \tau^{2\alpha k} E(x, r) \, . \tag{89}$$

Proof of Lemma 3.40. The lemma is proved by induction on k. Since it is clearly true for $k = 0$ we now assume that (88) and (89) hold for $0 \leq k \leq m-1$ and for some $m \in \mathbb{N}$. With Lemma 3.32 we obtain assuming that $\eta \leq \varepsilon(2L, \tau)$

$$E(x, \tau^m r) \leq C_*(2L) \tau^2 E(x, \tau^{m-1} r)$$

$$\leq \tau^{2\alpha} \tau^{2\alpha(m-1)} E(x, r)$$

$$= \tau^{2\alpha m} E(x, r) \, ,$$

hence (89) is valid for $k = m$.

To establish (88) with $k = m$ we first consider the case $q \geq 2$ and observe that

$$
\left|(\nabla u)_{x,\tau^{k+1}r} - (\nabla u)_{x,\tau^k r}\right| \leq \fint\limits_{B_{\tau^{k+1}r}(x)} \left|\nabla u - (\nabla u)_{x,\tau^k r}\right| \, dx
$$

$$
\leq \frac{1}{\tau^n} \fint\limits_{B_{\tau^k r}(x)} \left|\nabla u - (\nabla u)_{x,\tau^k r}\right| \, dx \qquad (90)
$$

$$
\leq \frac{1}{\tau^n} \left[E^+(x, \tau^k r)\right]^{\frac{1}{2}} .
$$

Thus, together with the assumptions (86) and (87) we arrive at

$$
\left|(\nabla u)_{x,\tau^m r}\right| \leq \sum_{k=0}^{m-1} \left|(\nabla u)_{x,\tau^{k+1}r} - (\nabla u)_{x,\tau^k r}\right| + \left|(\nabla u)_{x,r}\right|
$$

$$
\leq \frac{1}{\tau^n} \sum_{k=0}^{m-1} \left[E^+(x, \tau^k r)\right]^{\frac{1}{2}} + \left|(\nabla u)_{x,r}\right|
$$

$$
\leq L + \frac{1}{\tau^n} \eta^{\frac{1}{2}} \sum_{k=0}^{m-1} \tau^{\alpha k} \leq 2L
$$

if $\eta = \eta(L, \tau)$ is chosen sufficiently small.

It remains to consider the case $q < 2$. Here we additionally observe (following [CFM]) that Lemma 3.31, $i)$ and $vi)$ imply for any $B_\rho(x) \Subset B_R(x_0)$ satisfying $|(\nabla u)_{x,\rho}| \leq 2L$

$$
\fint\limits_{B_\rho(x)} \left|\nabla u - (\nabla u)_{x,\rho}\right| \, dx
$$

$$
\leq \frac{c}{\rho^n} \left\{ \int_{B_\rho(x) \cap [|\nabla u - (\nabla u)_{x,\rho}| \leq 1]} \left|\nabla u - (\nabla u)_{x,\rho}\right| \, dx \right.
$$

$$
\left. + \int_{B_\rho(x) \cap [|\nabla u - (\nabla u)_{x,\rho}| > 1]} \left|\nabla u - (\nabla u)_{x,\rho}\right| \, dx \right\} \qquad (91)
$$

$$
\leq \frac{c}{\rho^n} \left\{ \int_{B_\rho(x)} V\left(\nabla u - (\nabla u)_{x,\rho}\right) \, dx \right.
$$

$$
\left. + \int_{B_\rho(x)} V\left(\nabla u - (\nabla u)_{x,\rho}\right)^{\frac{2}{q}} \, dx \right\}
$$

$$
\leq c(2L) \left(\left[E^-(x, \rho)\right]^{\frac{1}{2}} + \left[E^-(x, \rho)\right]^{\frac{1}{q}} \right) .
$$

Hence, for $k \leq m - 1$ we obtain as a substitute for (90)

$$\left|(\nabla u)_{x,\tau^{k+1}r} - (\nabla u)_{x,\tau^{k}r}\right| \leq \frac{c(2L)}{\tau^{n}}\left([E(x,\rho)]^{\frac{1}{2}} + [E(x,\rho)]^{\frac{1}{q}}\right).$$

With this estimate the above arguments clearly extend to the case $q < 2$ and the lemma is proved. $\square$

Given Lemma 3.40, Theorem 3.27 follows in the case $q \geq 2$ from the definition of E and from the standard theory (compare, for example, [Gia1], Chapter 3, Theorem 1.3). If $q < 2$, then we have to apply (91) once again to obtain the conclusion. $\square$

3.5 Comparison with some known results

It is mentioned in the introduction of this chapter that the notion of (s, μ, q)-growth provides a unified and extended approach to the regularity theory for non-standard elliptic variational problems with superlinear growth. Let us shortly discuss this claim and briefly compare the main Theorems 3.11 and 3.27 with some of the basic references. We do not want to go into details and give a precise list of the assumptions supposed in the literature, we merely compare the most important hypotheses.

3.5.1 The scalar case

A.2. ANISOTROPIC POWER GROWTH

Consider the case $N = 1$ and assume that f is of (p, q)-growth as stated in (2) of the introduction. In this case, some of the most important results are due to Marcellini (see [Ma2]–[Ma4]), which – roughly speaking – read as

i) Assume that $u \in W^{1}_{q,loc}(\Omega)$ is a solution of the problem $(\mathcal{P})$. Then u is of class $C^{1,\alpha}(\Omega)$ if

$$q < p\frac{n}{n-2}.$$

ii) A solution $u \in W^{1}_{q,loc}(\Omega)$ exists if

$$q < p\frac{n+2}{n}.$$

If we recall Theorem 3.11, then we like to remark

ad i). The assumption $u \in W^{1}_{q,loc}(\Omega)$ corresponds to the case $s = q$. Hence, the (s, μ, q)-condition (14) of Theorem 3.11 can be written as

$$q < p + s\frac{2}{n} \quad \Leftrightarrow \quad q < p\frac{n}{n-2}.$$

ad ii). The existence result corresponds to the case $s = p$. Here we observe

$$q < p + s\frac{2}{n} \quad \Leftrightarrow \quad q < p\frac{n+2}{n}.$$

Thus, the above results can be interpreted as special cases of Theorem 3.11.

We already mentioned that Marcellini also includes the growth rate of the energy density f in the considerations of [Ma6] and [Ma7]. On one hand, this is done in a quite general formulation. On the other hand, he has to restrict to the study of problems with superquadratic growth conditions. Nevertheless, it seems to be worth mentioning that in the case of variational problems with (s, μ, q)-growth, $s \geq 2$, our (s, μ, q)-condition coincides with the set of assumptions which were introduced earlier by Marcellini from a quite different point of view.

B. GROWTH CONDITIONS INVOLVING N-FUNCTIONS

Again we consider the scalar case $N = 1$ where we now concentrate on variational integrands involving N-functions. The main contributions are

i) $C_{loc}^{1,\alpha}$-regularity of the solution in the case $f(Z) = |Z|\ln(1 + |Z|)$ (see [MS]).

ii) In the general setting of nearly linear growth conditions the solution to the problem $(\mathcal{P})$ is known to be of class $C^{1,\alpha}(\Omega)$ provided that

$$q < 2 - \mu + \frac{2}{n}.$$

This is proved in [FuM].

Clearly, these results are contained in Theorem 3.11.
ad i). Choose $\mu = 1$, $s = 1$, $q = 1 + \varepsilon$.
ad ii). Consider the (s, μ, q)-condition with $s = 1$, i.e.

$$q < 2 - \mu + s\frac{2}{n} \quad \Leftrightarrow \quad q < 2 - \mu + \frac{2}{n}.$$

3.5.2 The vectorial setting

A.2. ANISOTROPIC POWER GROWTH

In the vector-valued case there are only few results available for the (p, q)-situation. According to Passarelli Di Napoli and Siepe (see [PS]), we have partial regularity for the solution of the problem $(\mathcal{P})$ if we assume that

$$2 \leq p < q < \min\left\{p + 1, \frac{pn}{n-1}\right\}.$$

With the choice $2 \leq s = p$, $\mu = 2 - p$, this is a corollary of Theorem 3.27 (which, in particular, also covers the subquadratic case). However, we should emphasize the following restriction: in [PS] the parameter q is involved via

$$c_1 |Z|^p \le f(Z) \le c_2 \left(1 + |Z|^q\right).$$

We always choose q with regard to the second derivative of f, which is a stronger assumption. This was outlined in Remark 3.5, *iii*).

Let us finally mention the paper [AF4] which only partially fits into a comparison with Theorem 3.27 because of the very special structure taken as a hypothesis. We refer to [BF5] for a detailed discussion of examples where the structure of [AF4] leads to better results than stated above (see also Example 5.7 below).

B. GROWTH CONDITIONS INVOLVING N-FUNCTIONS

To our knowledge, the list of contributions on partial regularity in the case B.2 of the introduction shrinks to [FO]. In [FO] a theorem of type (2) is established in the following framework:

i) $q < 2$;

ii) $\mu < 4/n$;

iii) the integrand is balanced in some sense.

Again the results are completely covered as particular cases of Theorem 3.27 (compare the variant supposing (16)). Moreover – and this seems to be even more important – the (s, μ, q)-condition enables us to handle variational problems without a balanced structure.

3.6 Two-dimensional anisotropic variational problems

In Section 3.4 partial regularity results for vector-valued problems were established in the quite general setting of (s, μ, q)-growth. Now we restrict ourselves to anisotropic energy densities with (p, q)-growth. Even with this restriction and if p and q differ not too much, it seems to be unknown whether singular points can be excluded in the case of two dimensions. Here we briefly prove a theorem of this kind, to be precise we consider a bounded Lipschitz domain Ω in $\mathbb{R}^2$ and let $f \colon R^{2N} \to [0, \infty)$ denote a function of class $C^2(\mathbb{R}^{2N})$ such that

$$\lambda \left(1 + |Z|^2\right)^{\frac{p-2}{2}} |Y|^2 \le D^2 f(Z)(Y, Y) \le \Lambda \left(1 + |Z|^2\right)^{\frac{q-2}{2}} |Y|^2 \qquad (92)$$

holds for all $Z, Y \in \mathbb{R}^{2N}$ with positive constants λ, Λ. Here p and q are fixed exponents such that $1 < p < q < \infty$. From the discussion of Remark 3.5 we obtain the growth estimate

$$a|Z|^p - b \le f(Z) \le A \left(|Z|^q + 1\right) \quad \text{for all } Z \in \mathbb{R}^{2N} \qquad (93)$$

and for suitable constants a, b, $A > 0$. With the help of a lemma due to Frehse and Seregin ([FrS], Lemma 4.1) we are going to prove

Theorem 3.41. *Let f satisfy condition (92) and let u denote the solution of the problem (P), where the boundary data u_0 are supposed to be of class $W_p^1(\Omega; \mathbb{R}^N)$ such that $J[u_0] < \infty$. If $1 < p < q$ and if we additionally have*

$$q < 2p, \tag{94}$$

then $u \in C^{1,\alpha}(\Omega; \mathbb{R}^N)$ for any exponent $0 < \alpha < 1$.

Remark 3.42. *Note that the condition (94) formally coincides with the (s, μ, q)-condition (14), Section 3.3, choosing $s = p$, $\mu = 2 - p$.*

Example 3.43. *i) As a standard model we may take*

$$\int_\Omega \left(|\partial_1 u|^2 + (1 + |\partial_2 u|^2)^{\frac{q}{2}} \right) dx$$

for some q in the interval $[2, 4)$.
ii) Another example covered by Theorem 3.41 is given by

$$\int_\Omega \left((1 + |\nabla u|^2)^{\frac{1+\gamma}{2}} + |\partial_2 u|^2 \right) dx$$

with $\gamma \in (0, 1]$.

Proof of Theorem 3.41. The proof is organized in three steps, regularization, derivation of a starting inequality and a final application of the Frehse-Seregin Lemma.

Step 1. (Regularization) We refer to the regularization given in Section 3.4.1 with respect to the disc $B_R(0) \Subset \Omega$ (which of course can be assumed without loss of generality), hence we have Lemma 3.28. Note that $i)$ of Lemma 3.28 now can be replaced by (we again use the notation $u_\delta = u_{\delta(\varepsilon)}^\varepsilon$).

$$u_\delta \rightharpoonup u \quad \text{in } W_p^1(B_R(0); \mathbb{R}^N) \text{ as } \delta \to 0. \tag{95}$$

In fact, $i)$ and the growth estimate (93) immediately give (95).

Now we recall the a priori L^q-estimates of Section 3.3.2. Since we have the assumption (94), Lemma 3.17 is applicable (compare Remark 3.18) and gives a real number $t > q$ such that

$$\|\nabla u_\delta\|_{L^t(B_r(0); \mathbb{R}^{2N})} \leq c(r, R) < \infty \tag{96}$$

for any radius $r < R$ with a constant c independent of δ. In particular, u is in the space $W_{t, loc}^1(B_R(0); \mathbb{R}^N)$.

Step 2. (Starting inequality) Consider a disc $B_{2r}(x_0) \subset B_R(0)$ and let $\eta \in C_0^1(B_{2r}(x_0))$ denote a cut-off function such that $\eta \equiv 1$ on $B_r(x_0)$, $0 \leq \eta \leq 1$, and $|\nabla \eta| \leq c/r$. Following the proof of Lemma 3.19 we obtain

$$\int_{B_R(0)} \eta^2 D^2 f_\delta(\nabla u_\delta)(\partial_\gamma \nabla u_\delta, \partial_\gamma \nabla u_\delta) \, dx$$

$$\leq -2 \int_{B_R(0)} \eta D^2 f_\delta(\nabla u_\delta)(\partial_\gamma \nabla u_\delta, \partial_\gamma [u_\delta - Qx] \otimes \nabla \eta) \, dx \tag{97}$$

for any $Q \in \mathbb{R}^{2N}$. Next observe

$$\left| D^2 f_\delta(\nabla u_\delta)(\partial_\gamma \nabla u_\delta, \partial_\gamma [u_\delta - Qx] \otimes \nabla \eta) \right|$$

$$\leq \left(D^2 f_\delta(\nabla u_\delta)(\partial_\gamma \nabla u_\delta, \partial_\gamma \nabla u_\delta) \right)^{\frac{1}{2}}$$

$$\cdot \left(D^2 f_\delta(\nabla u_\delta)(\partial_\gamma [u_\delta - Qx] \otimes \nabla \eta, \partial_\gamma [u_\delta - Qx] \otimes \nabla \eta) \right)^{\frac{1}{2}}$$

and write

$$H_\delta = \left(D^2 f_\delta(\nabla u_\delta)(\partial_\gamma \nabla u_\delta, \partial_\gamma \nabla u_\delta) \right)^{\frac{1}{2}}.$$

Then, uniform local higher integrability and Lemma 3.19 show that H_δ is a function of class $L^2_{loc}(B_R(0))$. From (97) we get $\left(T_r(x_0) = B_{2r}(x_0) - B_r(x_0)\right)$

$$\int_{B_r(x_0)} H_\delta^2 \, dx \leq cr^{-1} \int_{T_r(x_0)} H_\delta \sqrt{|D^2 f_\delta(\nabla u_\delta)|} \, |\nabla u_\delta - Q| \, dx$$

$$\leq cr^{-1} \left(\int_{T_r(x_0)} H_\delta^2 \, dx \right)^{\frac{1}{2}} \tag{98}$$

$$\cdot \left(\int_{T_r(x_0)} |D^2 f_\delta(\nabla u_\delta)| \, |\nabla u_\delta - Q|^2 \, dx \right)^{\frac{1}{2}}.$$

From our assumption (92) and from the definition of f_δ we deduce

$$|D^2 f_\delta(\nabla u_\delta)| \, |\nabla u_\delta - Q|^2 \leq c(1 + \delta) \left(1 + |\nabla u_\delta|^2\right)^{\frac{q-2}{2}} |\nabla u_\delta - Q|^2.$$

Let us again consider the field $V(\xi) = (1 + |\xi|^2)^{\frac{q-2}{4}} \xi$, $\xi \in \mathbb{R}^{2N}$ – this time in the case $q \geq 2$ which is assumed from now on. Then, from [Gia1], p. 151, we infer

$$\left(1 + |\xi|^2\right)^{\frac{q-2}{2}} |\xi - Q|^2 \leq c|V(\xi) - V(Q)|^2, \tag{99}$$

and this is exactly the place where $q \geq 2$ is needed. For $q < 2$ the left-hand side of (99) has to be replaced by (see Lemma 3.31) $(1 + |\xi|^2 + |Q|^2)^{\frac{q-2}{2}} |\xi - Q|^2$ making the next calculations impossible. Now, using (99) and returning to (98), we arrive at

$$\int_{B_r(x_0)} H_\delta^2 \, dx \leq cr^{-1} \left(\int_{T_r(x_0)} H_\delta^2 \, dx \right)^{\frac{1}{2}}$$

$$\cdot \left(\int_{T_r(x_0)} |V(\nabla u_\delta) - V(Q)|^2 \, dx \right)^{\frac{1}{2}} \tag{100}$$

with a constant c which is not depending on r, x_0 and $\delta < 1$. Since V is a diffeomorphism of $\mathbb{R}^{2N}$, we may choose Q in such a way that

$$V(Q) = \fint_{T_r(x_0)} V(\nabla u_\delta) \, dx ,$$

which enables us to estimate the second integral on the right-hand side of (100) with the help of Sobolev-Poincaré's inequality, thus

$$\int_{B_r(x_0)} H_\delta^2 \, dx \leq cr^{-1} \left(\int_{T_r(x_0)} H_\delta^2 \, dx \right)^{\frac{1}{2}} \int_{T_r(x_0)} \left| \nabla (V(\nabla u_\delta)) \right| dx . \quad (101)$$

Note that the weak differentiability of $V(\nabla u_\delta)$ is already established in Lemma 3.14, II). Finally, we observe (inequality $*$ is discussed in Remark 3.44 below)

$$\left| \nabla (V(\nabla u_\delta)) \right| \overset{*}{\leq} c \left(1 + |\nabla u_\delta|^2 \right)^{\frac{q-2}{4}} |\nabla^2 u_\delta|$$

$$= c \left(1 + |\nabla u_\delta|^2 \right)^{\frac{p-2}{4}} |\nabla^2 u_\delta| \left(1 + |\nabla u_\delta|^2 \right)^{\frac{q-p}{4}}$$

$$\leq c \left(1 + |\nabla u_\delta|^2 \right)^{\frac{p-2}{4}} |\nabla^2 u_\delta| \left(1 + |\nabla u_\delta|^2 \right)^{\frac{p}{4}}$$

on account of $q < 2p$, and we may use the inequality (92) to get

$$\left(1 + |\nabla u_\delta|^2 \right)^{\frac{p-2}{4}} |\nabla^2 u_\delta| \leq cH_\delta .$$

Letting $h_\delta = \left(1 + |\nabla u_\delta|^2 \right)^{\frac{p}{4}}$, (101) implies

$$\int_{B_r(x_0)} H_\delta^2 \, dx \leq cr^{-1} \left(\int_{T_r(x_0)} H_\delta^2 \, dx \right)^{\frac{1}{2}} \int_{T_r(x_0)} H_\delta h_\delta \, dx , \quad (102)$$

being valid for any disc $B_{2r}(x_0) \subset B_R(0)$.

Remark 3.44. *The above inequality $*$ needs a technical comment. Let us write $V_\delta := V(\nabla u_\delta) = h_\delta^{\frac{q-2}{p}} \nabla u_\delta$. In Proposition 3.20 we have proved that $h_\delta \in W_{2,loc}^1(B_R(0))$ and that we have*

$$\nabla h_\delta = \frac{p}{2} |\nabla u_\delta| \left(1 + |\nabla u_\delta|^2 \right)^{\frac{p}{4}-1} \nabla |\nabla u_\delta| .$$

Since we consider the two-dimensional case $n = 2$ and on account of $h_\delta \geq 1$, we see that

$$h_\delta^{\frac{q-2}{p}} \in W_{t,loc}^1(B_R(0)) \quad \text{for any } t < 2,$$

and $u_\delta \in W_{2,loc}^2(B_R(0); \mathbb{R}^N)$ implies $\nabla u_\delta \in L_{loc}^s(B_R(0); \mathbb{R}^{2N})$ for any finite s. Observing also $h_\delta \in L_{loc}^s(B_R(0))$, $s < \infty$, we clearly have $V_\delta \in W_{1,loc}^1(B_R(0); \mathbb{R}^{2N})$ together with

$$\partial_\alpha V_\delta = \partial_\alpha \left(h_\delta^{\frac{q-2}{p}} \right) \nabla u_\delta + h_\delta^{\frac{q-2}{p}} \partial_\alpha \nabla u_\delta$$

$$= \frac{q-2}{p} h_\delta^{\frac{q-2}{p}-1} \partial_\alpha h_\delta \nabla u_\delta + h_\delta^{\frac{q-2}{p}} \partial_\alpha \nabla u_\delta .$$

Using the formula for $\partial_\alpha h_\delta$ we have proved $*$.

Step 3. (Application of the Frehse-Seregin Lemma) Inequality (102) exactly corresponds to the hypotheses of Lemma 4.1 in [FrS], and we get: for any $s \geq 1$ and for any compact subdomain ω of $B_R(0)$ there is a constant $K = K(\omega, s)$ such that

$$\int_{B_r(x_0)} H_\delta^2 \, dx \leq K |\ln r|^{-s} \tag{103}$$

is true for any disc $B_r(x_0) \Subset \omega$. Note that K also depends on $\|H_\delta\|_{L^2(\omega)}$ and $\|h_\delta\|_{W_2^1(\omega)}$ but on account of Lemma 3.19, Lemma 3.28 and Proposition 3.29 with the choice $\mu = 2 - p$, these quantities stay bounded uniformly with respect to δ. Let $G_\delta = (1 + |\nabla u_\delta|^2)^{\frac{p-2}{4}} \nabla u_\delta$. Recalling $h_\delta \in W_{2,loc}^1(B_R(0))$ we see that

$$\left(1 + |\nabla u_\delta|^2 \right)^{\frac{p-2}{4}} = h_\delta^{\frac{p-2}{p}} = h_\delta^{1-\frac{2}{p}}$$

belongs to the same function space. From Lemma 3.14 we know that u_δ is an element of $W_{2,loc}^2(B_R(0); \mathbb{R}^{2N})$, thus $G_\delta \in W_{1,loc}^1(B_R(0); \mathbb{R}^{2N})$, and for the derivative we get (recall Proposition 3.20) $|\nabla G_\delta| \leq c(1 + |\nabla u_\delta|^2)^{\frac{p-2}{4}} |\nabla^2 u_\delta|$ so that $|\nabla G_\delta|^2 \leq c H_\delta^2$. Therefore (103) implies

$$\int_{B_r(x_0)} |\nabla G_\delta|^2 \, dx \leq K |\ln r|^{-s}, \tag{104}$$

and if we choose $s > 2$ in (104), then the version of the Dirichlet-Growth Theorem given in [Fre], p.287, implies the continuity of G_δ on ω with modulus of continuity independent of δ. In Proposition 3.29, *iii*), we showed $\nabla u_\delta \to \nabla u$ almost everywhere on $B_R(0)$, therefore $G = (1 + |\nabla u|^2)^{\frac{p-2}{4}} \nabla u$ is a continuous function. Since $\xi \mapsto (1 + |\xi|^2)^{\frac{p-2}{4}} \xi$ is a homeomorphism (note that this field is proportional to the gradient of the strictly convex potential $(1 + |\xi|^2)^{\frac{p+2}{4}}$), we finally get continuity of ∇u. Thus, the criterion for regular points stated in Lemma 3.32 (which works in case $n = 2$ just under the condition (94)) is satisfied everywhere which proves the claim of Theorem 3.41 in case $q \geq 2$.

Let us now look at the case that (92) is valid with exponents $1 < p < q < 2$. But then (92) also holds for the new choice $q = 2$, i.e. we have

$$\lambda(1 + |Z|^2)^{\frac{p-2}{2}} |Y|^2 \leq D^2 f(Z)(Y,Y) \leq \Lambda |Y|^2,$$

and we may repeat all our calculations with q replaced by the exponent 2. Since $2 < 2p$, the appropriate version of condition (94) holds which gives the result. $\square$

Remark 3.45. *i) We like to remark that it is not necessary to refer to partial regularity results, a direct proof based on the inequality (102) can be obtained as in [FrS], Theorem 2.4.*

ii) In [BF6] it is shown in addition that Theorem 3.41 can be extended to handle also some limit cases like

$$f(Z) = |Z| \ln(1 + |Z|) + |z_2|^2 , \quad Z = (z_1, z_2) \in \mathbb{R}^{2N} ,$$

for which the assumption (94) is violated. By considering sufficiently regular boundary data (as done in the next chapters), we could show that minimizers are continuously differentiable functions. It remains an open question whether similar results are true for μ-elliptic variational integrands with $\mu > 1$.

Variational problems with linear growth: the case of μ-elliptic integrands

In this chapter we return to variational problems with linear growth but in contrast to Chapter 2 we study the variational problem $(\mathcal{P})$,

$$J[w] := \int_\Omega f(\nabla w) \, dx \to \min \quad \text{in } u_0 + \overset{\circ}{W}^1_1(\Omega; \mathbb{R}^N) \,, \qquad (\mathcal{P})$$

under more restrictive assumptions on the integrand f which might be sufficient for proving everywhere regularity of generalized minimizers in the scalar case $N = 1$ or in the vectorial case with additional structure.

The hypotheses formulated below are motivated by (the linear growth) Example 3.9, in particular by the condition $iii)$ of this example.

Assumption 4.1. *Throughout this chapter we consider variational integrands f satisfying the following set of hypotheses:*

$i)$ *f is of class $C^2(\mathbb{R}^{nN})$;*

$ii)$ *$|\nabla f(Z)| \leq A$;*

$iii)$ *$\lambda \left(1 + |Z|^2\right)^{-\frac{\mu}{2}} |Y|^2 \leq D^2 f(Z)(Y,Y) \leq \Lambda \left(1 + |Z|^2\right)^{-\frac{1}{2}} |Y|^2$.*

Here A, λ, Λ denote positive constants, $\mu > 1$ is some fixed exponent, and $ii)$, $iii)$ are valid for any choice of Z, $Y \in \mathbb{R}^{nN}$. An integrand f with the above properties is called μ-elliptic.

Remark 4.2. *Assumption 4.1 implies the following structure conditions.*

$i)$ *There are real numbers $\nu_1 > 0$ and ν_2 such that for all $Z \in \mathbb{R}^{nN}$*

$$\nabla f(Z) : Z \geq \nu_1 \left(1 + |Z|^2\right)^{\frac{1}{2}} - \nu_2 \,.$$

$ii)$ *The integrand f is of linear growth in the sense that for real numbers $\nu_3 > 0$, ν_4, $\nu_5 > 0$, ν_6 and for all $Z \in \mathbb{R}^{nN}$*

$$\nu_3 |Z| - \nu_4 \leq f(Z) \leq \nu_5 |Z| + \nu_6 \,.$$

iii) The integrand satisfies a balancing condition: there is a positive number ν_7 such that

$$|D^2 f(Z)||Z|^2 \leq \nu_7 \left(1 + f(Z)\right) \quad \text{holds for all } Z \in \mathbb{R}^{nN} .$$

Proof. ad *i*). Recall that we are interested in the variational problem $(\mathcal{P})$. We replace f by $\bar{f} \colon \mathbb{R}^{nN} \to \mathbb{R}$,

$$\bar{f}(Z) := f(Z) - \nabla f(0) : Z , \quad Z \in \mathbb{R}^{nN} .$$

An integration by parts gives a real number c such that we have for all $w \in u_0 + \overset{\circ}{W}{}^1_1(\Omega; \mathbb{R}^N)$

$$\bar{J}[w] := \int_\Omega f(\nabla w) \, \mathrm{d}x - \int_\Omega \nabla f(0) : \nabla w \, \mathrm{d}x = J[w] + c .$$

Thus minimizing sequences and generalized minimizers of J and $\bar{J}$, respectively, coincide, and without loss of generality $\nabla f(0) = 0$ can be assumed. This implies by Assumption 4.1, *iii*)

$$
\begin{aligned}
\nabla f(Z) : Z &= \int_0^1 \frac{d}{d\theta} \nabla f(\theta Z) : Z \, \mathrm{d}\theta \\
&= \int_0^1 D^2 f(\theta Z)(Z, Z) \, \mathrm{d}\theta \\
&\geq \lambda \int_0^1 \left(1 + \theta^2 |Z|^2\right)^{-\frac{\mu}{2}} |Z|^2 \, \mathrm{d}\theta \\
&= \lambda |Z| \int_0^{|Z|} \left(1 + \rho^2\right)^{-\frac{\mu}{2}} \, \mathrm{d}\rho ,
\end{aligned}
\tag{1}
$$

i.e. $\nabla f(Z) : Z$ is at least of linear growth and *i*) follows.

ad *ii*). The upper bound is immediate by Assumption 4.1, *ii*). Proving the left-hand inequality we observe that (1) gives $\nabla f(Z) : Z \geq 0$ for all $Z \in \mathbb{R}^{nN}$. Without loss of generality we additionally assume that $f(0) = 0$ to write (using *i*))

$$
\begin{aligned}
f(Z) &= \int_0^1 \frac{d}{d\theta} f(\theta Z) \, \mathrm{d}\theta \\
&\geq \int_{1/2}^1 \nabla f(\theta Z) : \theta Z \, \mathrm{d}\theta \\
&\geq \frac{1}{2} \left[\nu_1 \left(1 + \frac{|Z|^2}{4}\right)^{\frac{1}{2}} - \nu_2 \right],
\end{aligned}
$$

hence *ii*) is clear as well.

ad *iii*). This assertion follows from *ii*) and the right-hand side of Assumption 4.1, *iii*). $\square$

If $\mu < 1 + 2/n$ and if we have the structure conditions (11) and (13) of Section 3.3 in the vectorial setting, then Assumption 4.1 provides a regular class of variational integrands in the sense that generalized minimizers of the problem $(\mathcal{P})$ satisfy Theorem 1 and Theorem 3 of the introduction. Moreover, the elements of the set $\mathcal{M}$, i.e. generalized minimizers, merely differ by a constant. These results are established in Section 4.1.

The limitation $\mu < 1 + 2/n$ was already discussed in the introduction. On the other hand, the minimal surface integrand formally is μ-elliptic with $\mu = 3$. More precisely we have

Remark 4.3. *The minimal surface example $f(Z) = \sqrt{1 + |Z|^2}$ satisfies Assumption 4.1 with the limit exponent $\mu = 3$. However, there is much better information on account of the geometric structure of this example, in particular we have*

$$\frac{c_1}{\sqrt{1 + |Z|^2}} \left[|Y|^2 - \frac{(Y \cdot Z)^2}{1 + |Z|^2} \right] \leq D^2 f(Z)(Y, Y)$$

$$\leq \frac{c_2}{\sqrt{1 + |Z|^2}} \left[|Y|^2 - \frac{(Y \cdot Z)^2}{1 + |Z|^2} \right]$$

for all $Z, Y \in \mathbb{R}^n$ and with some real numbers c_1, c_2.

Given an integrand satisfying this condition, Ladyzhenskaya/Ural'tseva ([LU2]) and Giaquinta/Modica/Souček ([GMS1]) (see also [LU3]) then use Sobolev's inequality for functions defined on minimal hypersurfaces (compare [Mi] and [BGM]) as an essential tool for proving their regularity results.

Section 4.2 is devoted to the question, whether the limit $\mu = 3$ is of some relevance if the geometric structure condition is dropped. Here it turns out that we first have to discuss some examples.

Then, with some natural boundedness assumption (which can be deduced from a suitable maximum principle), we continue by considering the vector-valued setting: a generalized minimizer u^* of class $W_1^1(\Omega; \mathbb{R}^N)$ is found. Moreover, u^* uniquely (up to a constant) determines the solutions of the following problem:

$$\int_\Omega f(\nabla w) \, dx + \int_{\partial \Omega} f_\infty \big((u_0 - w) \otimes \nu \big) \, d\mathcal{H}^{n-1} \to \min \quad \text{in} \quad W_1^1(\Omega; \mathbb{R}^N) . \quad (\mathcal{P}')$$

Here f_∞ denotes the recession function and ν is the unit outward normal to $\partial \Omega$. Note that these results neither depend on a geometric structure condition nor on the assumption $f(Z) = g(|Z|^2)$.

If – as a substitute for the geometric structure – $\mu < 3$ is assumed, then the uniqueness of generalized BV-minimizers up to a constant as well as Theorem 1 and Theorem 3 of the introduction are also established in Section 4.2. This substantially improves the results of Section 4.1 for locally bounded generalized solutions (in the sense of Assumption 4.11 or Remark 4.12).

In Section 4.3 it is shown that boundedness conditions are superfluous while discussing two-dimensional problems. Here we again have to rely on the structure $f(Z) = g(|Z|^2)$ in the case of vector-valued problems.

A complete picture is obtained by the sharpness result of Section 4.4. We benefit from ideas of [GMS1] and show that no smoothness results can be expected if the ellipticity exponent satisfies $\mu > 3$. In view of this x-dependent example we already have included (as a model case) Section 4.2.2.2 on a smooth x-dependence.

4.1 The case $\mu < 1 + 2/n$

In this section we are going to establish our first result on full regularity and uniqueness of generalized minimizers of the problem $(\mathcal{P})$, where we always assume that $u_0 \in W^1_p(\Omega; \mathbb{R}^N)$ for some $p > 1$ (by considering a suitable approximation it is also possible to handle the limit case $p = 1$, see Remark 2.5). Recall the notion of the set of generalized minimizers

$$\mathcal{M} = \left\{ u \in BV(\Omega; \mathbb{R}^N) : u \text{ is the } L^1\text{-limit of a } J\text{-minimizing sequence} \right.$$

$$\left. \text{from } u_0 + \overset{\circ}{W}{}^1_1(\Omega; \mathbb{R}^N) \right\} .$$

Moreover, if $N > 1$, we want to benefit from the above mentioned structure, thus we assume: there are constants K and $0 < \alpha < 1$ such that for all Z, $\tilde{Z} \in \mathbb{R}^{nN}$

$$f(Z) = g(|Z|^2) , \quad g \in C^2([0,\infty); [0,\infty)) , \tag{2}$$

$$\left| D^2 f(Z) - D^2 f(\tilde{Z}) \right| \leq K |Z - \tilde{Z}|^\alpha . \tag{3}$$

Now we formulate

Theorem 4.4. *Consider a variational integrand f satisfying Assumption 4.1 with*

$$\mu < 1 + \frac{2}{n} . \tag{4}$$

Suppose in addition that (2) and (3) are valid in the vectorial setting $N > 1$. Then the following assertions are true.

i) *Any generalized minimizer $u \in \mathcal{M}$ is in the space $C^{1,\alpha}(\Omega; \mathbb{R}^N)$ for any $0 < \alpha < 1$.*

ii) *For u, $v \in \mathcal{M}$ we have $\nabla u = \nabla v$, i.e. generalized minimizers are unique up to a constant.*

Remark 4.5. *In [BF3] the limit case $n = 2$, $\mu = 2$ could be included with the help of John-Nirenberg estimates (see [GT], Theorem 7.21, p. 166). Here we do not follow these ideas since the technique outlined in Section 4.3 provides much stronger results.*

Remark 4.6. *Uniqueness up to constants as stated in Theorem 4.4, ii), is exactly what we expect recalling the known results in the minimal surface case (see the monograph of Giusti [Giu2]).*

The *Proof of Theorem 4.4* relies on the a priori estimates given in Section 3.3. As a second ingredient the uniqueness Theorem A.9 is needed.

4.1.1 Regularization

Let the assumptions of Theorem 4.4 hold. We fix some real number $1 < q < 2$ satisfying $q < p$ and in addition for $n \geq 3$ (recall (4))

$$q < (2 - \mu)\frac{n}{n-2} . \tag{5}$$

Now we follow exactly the lines of Section 3.3, define for any $0 < \delta < 1$

$$J_\delta[w] := \delta \int_\Omega \left(1 + |\nabla w|^2\right)^{\frac{q}{2}} \mathrm{d}x + J[w] , \quad w \in u_0 + \overset{\circ}{W}{}^1_q(\Omega; \mathbb{R}^N) ,$$

and denote by u_δ the unique solution of

$$\text{to minimize } J_\delta[w] \text{ in the class } u_0 + \overset{\circ}{W}{}^1_q(\Omega; \mathbb{R}^N). \tag{$\mathcal{P}_\delta$}$$

Thus, letting $f_\delta(\cdot) = \delta\left(1 + |\cdot|^2\right)^{\frac{q}{2}} + f(\cdot)$, we obtain

$$\int_\Omega \nabla f_\delta(\nabla u_\delta) : \nabla \varphi \, \mathrm{d}x = 0 \quad \text{for all } \varphi \in \overset{\circ}{W}{}^1_q(\Omega; \mathbb{R}^N) . \tag{6}$$

Remark 4.7.

  i) *It should be emphasized that this "approximative part" of the proof completely coincides with the situation of Section 3.3.*

 ii) *In particular, on account of (4) and (5), uniform local a priori gradient estimates are deduced from Theorem 3.16. Here we have to recall that the situation is a "balanced" one in the sense of Remark 4.2, iii).*

4.1.2 Some remarks on the dual problem

We shortly pass to the dual variational problem defined in Section 2.1.1. This will provide some auxiliary results needed in the following.

  The regularization at hand slightly differs from the one given in Section 2.1.2 where the case $q = 2$ was considered. We now let

$$\tau_\delta := \nabla f(\nabla u_\delta) , \quad \sigma_\delta := \delta X_\delta + \tau_\delta = \nabla f_\delta(\nabla u_\delta) ,$$
$$X_\delta := q\left(1 + |\nabla u_\delta|^2\right)^{\frac{q-2}{2}} \nabla u_\delta .$$

Note that $\sigma_\delta \in W^1_{2,loc}(\Omega; \mathbb{R}^{nN})$ which, for $N > 1$, follows from [AF3], Proposition 2.7. The arguments outlined in the second part of Section 2.1.2 now reads as: since $J_\delta[u_\delta] \leq J_\delta[u_0] \leq J_1[u_0]$, there exist positive numbers c_1, c_2, c_3 such that

$$\delta \int_\Omega \left(1 + |\nabla u_\delta|^2\right)^{\frac{q}{2}} dx \leq c_1 , \quad \int_\Omega f(\nabla u_\delta) \, dx \leq c_2 , \quad \|\tau_\delta\|_\infty \leq c_3 . \qquad (7)$$

The first inequality implies

$$\left\| \delta^{\frac{q-1}{q}} X_\delta \right\|_{L^{\frac{q}{q-1}}(\Omega; \mathbb{R}^{nN})} \leq c , \quad \text{hence} \quad \delta X_\delta \rightharpoonup 0 \text{ in } L^{\frac{q}{q-1}}(\Omega; \mathbb{R}^{nN}) \qquad (8)$$

as $\delta \to 0$. Here and in the following we (as usual) pass to subsequences if necessary. By (7) and (8) we find $\sigma \in L^{\frac{q}{q-1}}(\Omega; \mathbb{R}^{nN})$ such that

$$\tau_\delta , \sigma_\delta \rightharpoonup : \sigma \quad \text{in } L^{\frac{q}{q-1}}(\Omega; \mathbb{R}^{nM}) \text{ as } \delta \to 0 . \qquad (9)$$

Lemma 4.8.

  i) *The limit σ given in (9) is admissible in the sense that* div $\sigma = 0$.

  ii) *Assume that σ is given as above. Then σ maximizes the dual variational problem $(\mathcal{P}^*)$.*

  iii) *The unique maximizer σ is of class $W^1_{2,loc}(\Omega; \mathbb{R}^{nN})$.*

  iv) *The sequence $\{u_\delta\}$ is a J-minimizing sequence.*

Proof. The claim div $\sigma = 0$ again is a direct consequence of div $\sigma_\delta = 0$. To prove the second assertion, recall the duality relation

$$\tau_\delta : \nabla u_\delta - f^*(\tau_\delta) = f(\nabla u_\delta) ,$$

which, together with the definition of σ_δ and with div $\sigma_\delta = 0$, gives

$$J_\delta[u_\delta] = \delta \int_\Omega \left(1 + |\nabla u_\delta|^2\right)^{\frac{q}{2}} dx + \int_\Omega \left(\sigma_\delta : \nabla u_0 - f^*(\tau_\delta)\right) dx$$
$$-\delta q \int_\Omega \left(1 + |\nabla u_\delta|^2\right)^{\frac{q-2}{2}} |\nabla u_\delta|^2 \, dx .$$

This yields for all $\varkappa \in L^\infty(\Omega; \mathbb{R}^{nN})$

$$R[\varkappa] \leq \inf_{u \in u_0 + \overset{\circ}{W}^1_1(\Omega; \mathbb{R}^N)} J[u] \leq J[u_\delta] \leq J_\delta[u_\delta]$$
$$= \int_\Omega \left(\tau_\delta : \nabla u_0 - f^*(\tau_\delta)\right) dx + \delta \int_\Omega X_\delta : \nabla u_0 \, dx \qquad (10)$$
$$+(1-q)\delta \int_\Omega \left(1 + |\nabla u_\delta|^2\right)^{\frac{q}{2}} dx + \delta q \int_\Omega \left(1 + |\nabla u_\delta|^2\right)^{\frac{q-2}{2}} dx ,$$

and, passing to the limit $\delta \to 0$, the second and the last integral on the right-hand side both vanish according to (8) and since $q < 2$. As in Section 2.1.2, lower semicontinuity of $\int_{\Omega} f^*(\cdot)\,\mathrm{d}x$ with respect to weak-$*$ convergence proves the claim $R[\varkappa] \leq R[\sigma]$ as well as

$$\delta \int_{\Omega} \left(1 + |\nabla u_{\delta}|^2\right)^{\frac{q}{2}} \mathrm{d}x \to 0 \quad \text{in the limit } \delta \to 0. \tag{11}$$

Note that the minimizing property of $\{u_{\delta}\}$ is established with (10). Finally, the arguments leading to local W_2^1-regularity are completely the same as given Section 2.1.3 (recall that we have the starting inequality of Lemma 3.19). However, we do not exploit iii) in the following. $\quad\square$

4.1.3 Proof of Theorem 4.4

With Remark 4.7, ii), and Lemma 4.8, iv), it is obvious that each L^1-cluster point u^* of the sequence $\{u_{\delta}\}$ is a locally Lipschitz generalized minimizer in the class $\mathcal{M}$. In particular, a generalized minimizer with vanishing singular part of the derivative is found and, following Section 2.3.1 or Appendix A.1, the Euler equation

$$\int_{\Omega} \nabla f(\nabla u^*) : \nabla\varphi \, \mathrm{d}x = 0 \quad \text{for all } \varphi \in C_0^1(\Omega; \mathbb{R}^N) \tag{12}$$

is verified. Hence, the standard arguments as sketched at the end of Section 3.3.4 yield Hölder continuity of the derivatives and so far it is proved that

Proposition 4.9. *Let the assumptions of Theorem 4.4 hold. If u^* denotes a L^1-cluster point of the sequence $\{u_{\delta}\}$, then $u^* \in \mathcal{M}$ and u^* is of class $C^{1,\alpha}(\Omega; \mathbb{R}^N)$ for any $0 < \alpha < 1$.*

Now the arguments of Section 2.3.2 are modified to obtain

Proposition 4.10. *With the assumptions of Theorem 4.4 and with u^*, σ as above we have*

$$\sigma = \nabla f(\nabla u^*).$$

In particular, for any $0 < \alpha < 1$, the unique dual solution σ is of class $C^{0,\alpha}(\Omega; \mathbb{R}^{nN})$.

Proof. Recalling the Euler equation (12), we choose $\varphi = \eta^2 (u_{\delta} - u^*)$, $\eta \in C_0^1(\Omega)$, $0 \leq \eta \leq 1$. Then, together with (6), the counterpart of (29), Section 2.3.2, is established:

$$\int_{\Omega} \eta^2 \left(\nabla f(\nabla u_{\delta}) - \nabla f(\nabla u^*)\right) : (\nabla u_{\delta} - \nabla u^*) \, \mathrm{d}x$$

$$+ \delta \int_{\Omega} \eta^2 X_{\delta} : (\nabla u_{\delta} - \nabla u^*) \, \mathrm{d}x$$

$$= -2 \int_{\Omega} \sigma_{\delta} : ([u_{\delta} - u^*] \otimes \nabla\eta) \, \eta \, \mathrm{d}x$$

$$+ 2 \int_{\Omega} \nabla f(\nabla u^*) : ([u_{\delta} - u^*] \otimes \nabla\eta) \, \eta \, \mathrm{d}x.$$

Clearly the second integral on the right-hand side vanishes as $\delta \to 0$ and by (8), (11) this is also true for the second one on the left-hand side. Since the definition of σ_δ gives the same result for the first integral on the right-hand side, it is proved that

$$\lim_{\delta \downarrow 0} \int_\Omega \left(\nabla f(\nabla u_\delta) - \nabla f(\nabla u^*) \right) : (\nabla u_\delta - \nabla u^*) \eta^2 \, dx = 0 \, .$$

Finally, the proof is completed exactly as in Section 2.3.2.   □

Now that Proposition 4.9 and Proposition 4.10 are established, Theorem 4.4 follows as a corollary of the uniqueness result given in Theorem A.9.   □

4.2 Bounded generalized solutions

In this section we cover the whole scale of μ-elliptic integrands with linear growth (as introduced in Assumption 4.1) up to $\mu = 3$. As discussed in Remark 4.3, the case $\mu = 3$ is the limit case induced by the minimal surface example, where one benefits from the geometric structure to obtain a priori gradient estimates. On account of the lack of this structure, in the general situation we have to rely on an additional assumption which, nevertheless, is a very natural one: the boundary values u_0 are supposed to be of class $L^\infty(\Omega; \mathbb{R}^N)$ (following Remark 2.5, without loss of generality $u_0 \in L^\infty \cap W_2^1(\Omega; \mathbb{R}^N)$). Moreover, we suppose that a maximum principle is valid for the regularization. Note that in contrast to the previous section we again consider the standard regularization introduced in Section 2.1.2. Furthermore, the expression "bounded generalized solutions" is meant in the sense of "uniformly bounded regularizations".

Assumption 4.11. *Let u_δ denote the unique minimizer of*

$$J_\delta[w] := \frac{\delta}{2} \int_\Omega |\nabla w|^2 \, dx + J[w] \, , \qquad w \in u_0 + \overset{\circ}{W}{}_2^1(\Omega; \mathbb{R}^N) \, ,$$

$\delta \in (0, 1)$. *Then there is a real number M, independent of δ, such that*

$$\|u_\delta\|_{L^\infty(\Omega;\mathbb{R}^N)} \le M \|u_0\|_{L^\infty(\Omega;\mathbb{R}^N)} \, .$$

Remark 4.12.

 i) *Alternatively, Assumption 4.11 may be replaced by*

$$\|u_\delta\|_{L^\infty_{loc}(\Omega;\mathbb{R}^N)} \le K$$

 for some real number K not depending on δ. In this case no restriction on the boundary values is needed.

 ii) *In Section 5.1 the reader will find a formulation which seems to be quite natural for obtaining the convex hull property in the case $N > 1$ (compare (5) of Assumption 5.1 together with Remark 5.2, i)). Of course the convex hull property gives Assumption 4.11.*

Remark 4.13. *There are a lot of contributions on the boundedness of solutions of variational problems. Let us mention [Ta] in the scalar case, a maximum principle for $N > 1$ is given in [DLM]. Let us also remark that in the case of non-standard growth conditions, a boundedness assumption serves as an important tool in [Ch] and [ELM2] (see Chapter 5).*

As outlined in the introduction, given Assumption 4.11 we do not have to differentiate the Euler equation – avoiding the use of Sobolev's inequality – in order to obtain uniform local higher integrability of the gradients. This removes the first restriction on μ and gives an interesting existence result, even in the limit case $\mu = 3$.

Moreover, if $\mu < 3$, then integrability is improved up to any exponent. This information finally enables us to carry out a DeGiorgi-type technique. Note that, in spite of the strong higher integrability results, this modification is non-trivial for the range of μ under consideration: for instance, let us consider the proof of Lemma 3.25 and assume that $\mu > 2$. In this case we have to choose $\beta > q/2$ which, on the other hand, has to be excluded proving Lemma 3.25. Furthermore, observe that the conditions (14) and (16), Theorem 3.11, completely break down if $\mu > 2$. Even the two-dimensional considerations strongly depend on the assumption $\mu < 2$.

Now let us give a precise formulation of the main results of this section.

Theorem 4.14. *If $N \geq 1$ and if Assumption 4.1 and Assumption 4.11 are valid with $\mu = 3$, then there exists a generalized minimizer $u^* \in \mathcal{M}$ such that*

i) We have $\nabla^s u^* \equiv 0$, hence $\nabla u^* \equiv \nabla^a u^*$.

ii) For any $\Omega' \Subset \Omega$ there is a constant $c(\Omega')$ satisfying

$$\int_{\Omega'} |\nabla u^*| \ln^2 \left(1 + |\nabla u^*|^2\right) \, dx \leq c(\Omega') < \infty \,.$$

iii) The particular minimizer u^* is of class $W_1^1(\Omega; \mathbb{R}^N)$ and (up to a constant) the unique solution of the variational problem

$$\int_\Omega f(\nabla w) \, dx + \int_{\partial\Omega} f_\infty \left((u_0 - w) \otimes \nu\right) \, d\mathcal{H}^{n-1} \to \min \quad \text{in } W_1^1(\Omega; \mathbb{R}^N) \,,$$

where f_∞ denotes the recession function and where ν is the unit outward normal to $\partial\Omega$.

Remark 4.15. *It should be emphasized that no additional structure condition is needed in the above theorem to handle the vectorial setting.*

A slightly stronger ellipticity condition yields:

Theorem 4.16. *Suppose that Assumption 4.1 and Assumption 4.11 are valid with $\mu < 3$. In the case $N > 1$ we additionally assume (2) and (3) of Section 4.1. Then we have*

i) *Each generalized minimizer $u \in \mathcal{M}$ is in the space $C^{1,\alpha}(\Omega; \mathbb{R}^N)$ for any $0 < \alpha < 1$.*

ii) *For u, $v \in \mathcal{M}$ we have $\nabla u = \nabla v$, i.e. generalized minimizers are unique up to a constant.*

Before we are going to prove these theorems let us briefly discuss some examples.

Example 4.17.

i) *Consider the μ-elliptic linear growth function Φ which was introduced in Example 3.9. Suppose that $N = 1$, $1 < \mu < 3$ and fix Z, $Y \in \mathbb{R}^n$ satisfying $Z = \lambda Y$, $\lambda \in \mathbb{R}$ sufficiently large. Then*

$$\frac{1}{\sqrt{1 + |Z|^2}} \left[|Y|^2 - \frac{(Y \cdot Z)^2}{1 + |Z|^2} \right] = (1 + |Z|^2)^{-\frac{3}{2}} |Y|^2$$

$$< (1 + |Z|^2)^{-\frac{\mu}{2}} |Y|^2$$

$$\leq cD^2\Phi(Z)(Y, Y) ,$$

hence Φ is not of minimal surface structure: we do not have the upper bound given in Remark 4.3.

ii) *In the case $\mu = 1$, at least for large $|Z|$, the function Φ behaves like $|Z| \ln(1 + |Z|)$.*
If $\mu = 2$, then we have the representation

$$\Phi(Z) = |Z| \arctan|Z| - \frac{1}{2} \ln \left(1 + |Z|^2 \right) .$$

In the limit case $\mu = 3$ it is easy to perform the integrations with the result

$$\Phi(Z) = \sqrt{1 + |Z|^2} ,$$

hence the functions $\Phi = \Phi_\mu$ provide a one-parameter family connecting our logarithmic example with the minimal surface integrand.

iii) *On account of the above observation we need to give some examples with the limit ellipticity $\mu = 3$, with linear growth and which are not of minimal surface structure.*
A quite technical example can be constructed analogous to Step 1 leading to Example 3.7: consider the function Φ for fixed $1 < \mu < 3$ and "destroy" the ellipticity. Then, on one hand, the upper bound being valid in the minimal surface case does not hold (see i)). On the other hand, we have degenerate ellipticity. Finally, add some minimal surface part.
The reason why we need this complicated construction is the following: the Ansatz $f(Z) = g(|Z|)$ automatically leads (more or less) to the minimal

surface structure. In particular, if $Z \perp Y \in \mathbb{R}^n$ and if f is of linear growth, hence $g'(t) \to c$ as $t \to +\infty$, then

$$D^2 f(Z)(Y,Y) = \frac{g'(|Z|)}{|Z|}|Y|^2 \approx \left(1+|Z|^2\right)^{-\frac{1}{2}}|Y|^2$$

if $|Z|$ is sufficiently large.

Nevertheless, there exists a natural class of examples where the above structure is lost: the idea is to replace $|Z|$ by the distance to a convex set. This idea provides a variety of interesting energy densities. Let us just sketch a very easy example in the case $N = 1$, $n = 2$, $Z = (z_1, z_2)$. Denote by C the upper unit half disc, i.e. $C = \{Z : |Z| < 1, z_2 > 0\}$ (for the sake of simplicity we neglect a smoothing procedure at the edges). Note that the distance function $\rho(Z) := \operatorname{dist}(Z, C)$ coincides (up to the constant 1) in the upper half plane (for $|Z| > 1$) with $|Z|$. Now let

$$f(Z) = \sqrt{1 + \rho^2(Z)}\ .$$

We are mainly interested in the points $Z = (0, z_2)$, $z_2 < 0$, $|z_2| \gg 1$: it evident that in this case

$$D^2 f\big((0, z_2)\big)(e_1, e_1) = 0\ ,$$

$$D^2 f\big((0, z_2)\big)(e_2, e_2) = \left(1 + |\rho^2\big((0, z_2)\big)|\right)^{-\frac{3}{2}} = \left(1 + |Z|^2\right)^{-\frac{3}{2}}\ ,$$

where e_i, $i = 1, 2$, denotes the ith unit coordinate vector. In particular, we observe that the minimal surface structure is completely destroyed on account of the degeneracy of C. This of course induces degeneracy of f as well. The first way of obtaining μ-elliptic integrands with linear growth evidently is to change the geometry in a suitable way. We prefer a simple and more anisotropic idea: let (for $|Z| > 1$)

$$\tau\big((z_1, z_2)\big) = \frac{1}{2} z_1 + \rho\big((z_1, z_2)\big)\ ,$$

$$\tilde{f}\big((z_1, z_2)\big) = \sqrt{1 + \tau^2\big((z_1, z_2)\big)}\ .$$

Then there is a positive constant c such that

$$c^{-1}|Z| \le \tau\big((z_1, z_2)\big) \le c|Z|$$

for all $|Z|$ sufficiently large, and if $z_2 < 0$, $|z_2| \gg 1$, then we obtain in both coordinate directions e_i, $i = 1, 2$, with suitable constants c_i

$$D^2 \tilde{f}\big((0, z_2)\big)(e_i, e_i) = c_i \left(1 + |\tau\big((0, z_2)\big)|^2\right)^{-\frac{3}{2}}\ .$$

Summarizing the properties of $\tilde{f}$ we see that this function is of linear growth and satisfies the μ-ellipticity condition with the limit exponent $\mu =$

3. *Moreover, $\tilde{f}$ does not satisfy the minimal surface ellipticity condition from Remark 4.3 and there is no chance to get something analogous: given the points $(0, z_2)$ as above, both eigenvalues of $D^2 \tilde{f}((0, z_2))$ grow like $(1 + |Z|^2)^{-3/2}$.*

4.2.1 Regularization

In contrast to the previous chapter and in contrast to Section 4.1.1, no particular choice of the power of the regularization is induced by the variational problem, hence we consider the regularization as given in Assumption 4.11. Once again note that we have (letting $f_\delta := \frac{\delta}{2} | \cdot |^2 + f$) the uniform bound

$$\int_\Omega f_\delta(\nabla u_\delta) \, dx \leq c \tag{13}$$

for some real number c as well as the Euler equation

$$\int_\Omega \nabla f_\delta(\nabla u_\delta) : \nabla \varphi \, dx = 0 \quad \text{for all } \varphi \in \overset{\circ}{W}^1_2(\Omega; \mathbb{R}^N) . \tag{14}$$

As in Section 2.1.2 we denote $\sigma_\delta = \nabla f_\delta(\nabla u_\delta)$ and assume on account of (13) that $\sigma_\delta \rightharpoonup \sigma$ in $L^2(\Omega; \mathbb{R}^{nN})$ as $\delta \to 0$.

Remark 4.18. *Recall that the following claims were proved in Section 2.1.2.*

  i) *The sequence $\{u_\delta\}$ is a J-minimizing sequence. Hence, the L^1-cluster points of $\{u_\delta\}$ provide generalized minimizers $u^* \in \mathcal{M}$.*
  ii) *The limit σ of the sequence $\{\sigma_\delta\}$ maximizes the dual variational problem $(\mathcal{P}^*)$ which was introduced in Section 2.1.1.*

Next, Lemma 3.19 and Lemma 3.24 have to be generalized.

Lemma 4.19. *Suppose that Assumption 4.1 is true and that we have the structure condition (2), Section 4.1, in the case $N > 1$.*

  i) *There is a real number $c > 0$ such that for any $s \geq 0$, for all $\eta \in C^\infty_0(\Omega)$, $0 \leq \eta \leq 1$ and for any $\delta \in (0, 1)$*

$$\int_\Omega D^2 f_\delta(\nabla u_\delta)(\partial_\gamma \nabla u_\delta, \partial_\gamma \nabla u_\delta) \Gamma^s_\delta \eta^2 \, dx$$

$$\leq c \int_\Omega D^2 f_\delta(\nabla u_\delta)(\partial_\gamma u_\delta \otimes \nabla \eta, \partial_\gamma u_\delta \otimes \nabla \eta) \Gamma^s_\delta \, dx ,$$

  where we have set $\Gamma_\delta = 1 + |\nabla u_\delta|^2$.
  ii) *In contrast to Section 3.3.3 we denote in the following*

$$A(k, r) = A_\delta(k, r) = \{x \in B_r(x_0) : \Gamma_\delta > k\} , \quad k > 0 .$$

Then there is a real number $c > 0$, independent of δ, such that for all $\eta \in C_0^\infty(B_r(x_0))$, $0 \le \eta \le 1$ and for any $\delta \in (0,1)$

$$\int_{A(k,r)} \Gamma_\delta^{-\frac{\mu}{2}} |\nabla \Gamma_\delta|^2 \eta^2 \, dx$$

$$\le c \int_{A(k,r)} D^2 f_\delta(\nabla u_\delta)(e_j \otimes \nabla \eta, e_j \otimes \nabla \eta)(\Gamma_\delta - k)^2 \, dx .$$

Here e_j denotes the jth coordinate vector.

Remark 4.20. *Following the proof of Lemma 4.19 we see that the structure condition is not needed to establish the first claim in the case $s = 0$ (compare Lemma 3.19).*

Proof of Lemma 4.19. ad i). We have already seen at the beginning of Section 2.1.3 that the Euler equation (14) yields

$$\int_\Omega D^2 f_\delta(\nabla u_\delta)(\partial_\gamma \nabla u_\delta, \nabla \varphi) \, dx = 0 \quad \text{for all } \varphi \in C_0^\infty(\Omega; \mathbb{R}^N) , \qquad (15)$$

hence, using standard approximation arguments, (15) follows for all $\varphi \in W_2^1(\Omega; \mathbb{R}^N)$ which are compactly supported in Ω. Next we cite [LU1], Chapter 4, Theorem 5.2, in the scalar case and [Uh] (we may also refer to [GiaM], Theorem 3.1) if $N > 1$ to see that $u_\delta \in W_{\infty,loc}^1(\Omega; \mathbb{R}^N)$. As a consequence, $\varphi = \eta^2 \, \partial_\gamma u_\delta \, \Gamma_\delta^s$ (with η given above) is admissible in (15) (recall the product and the chain rule for Sobolev functions). Summarizing the results we arrive at

$$\int_\Omega D^2 f_\delta(\nabla u_\delta)(\partial_\gamma \nabla u_\delta, \partial_\gamma \nabla u_\delta) \, \Gamma_\delta^s \eta^2 \, dx$$

$$+ s \int_\Omega D^2 f_\delta(\nabla u_\delta)(\partial_\gamma \nabla u_\delta, \partial_\gamma u_\delta \otimes \nabla \Gamma_\delta) \, \Gamma_\delta^{s-1} \eta^2 \, dx \qquad (16)$$

$$= -2 \int_\Omega D^2 f_\delta(\nabla u_\delta)(\partial_\gamma \nabla u_\delta, \partial_\gamma u_\delta \otimes \nabla \eta) \, \eta \, \Gamma_\delta^s \, dx .$$

In the scalar case $N = 1$ the second integral on the left-hand side can be neglected on account of

$$D^2 f_\delta(\nabla u_\delta)(\partial_\gamma \nabla u_\delta, \partial_\gamma u_\delta \otimes \nabla \Gamma_\delta) = \frac{1}{2} D^2 f_\delta(\nabla u_\delta)(\nabla \Gamma_\delta, \nabla \Gamma_\delta) \ge 0 \text{ a.e.}$$

In the vectorial setting $N > 1$ we first consider the case $s = 0$. Then the second term on the left-hand side trivially vanishes without any additional assumption (compare Remark 4.20). If $s > 0$, then the structure condition is needed: given a weakly differentiable function $\psi \colon \Omega \to \mathbb{R}$ and letting $f_\delta(Z) = g_\delta(|Z|^2)$ we have proved in (32) of Section 3.3.3 that almost everywhere

$$D^2 f_\delta(\nabla u_\delta)(\partial_\gamma \nabla u_\delta, \partial_\gamma u_\delta \otimes \nabla \psi)$$
$$= \frac{1}{2} D^2 f_\delta(\nabla u_\delta)(e_j \otimes \nabla \psi, e_j \otimes \nabla \Gamma_\delta) \, . \tag{17}$$

Choosing $\psi = \Gamma_\delta$ we see that in the vectorial setting the second integral on the left-hand side of (16) is non-negative as well. In any case we obtain for a given $\varepsilon > 0$

$$\int_\Omega D^2 f_\delta(\nabla u_\delta)(\partial_\gamma \nabla u_\delta, \partial_\gamma \nabla u_\delta) \Gamma_\delta^s \eta^2 \, dx$$

$$\leq c \int_\Omega \left[D^2 f_\delta(\nabla u_\delta)(\partial_\gamma \nabla u_\delta, \partial_\gamma \nabla u_\delta) \right]^{\frac{1}{2}} \eta \Gamma_\delta^{\frac{s}{2}}$$

$$\cdot \left[D^2 f_\delta(\nabla u_\delta)(\partial_\gamma u_\delta \otimes \nabla \eta, \partial_\gamma u_\delta \otimes \nabla \eta) \right]^{\frac{1}{2}} \Gamma_\delta^{\frac{s}{2}} \, dx$$

$$\leq c \left\{ \varepsilon \int_\Omega D^2 f_\delta(\nabla u_\delta)(\partial_\gamma \nabla u_\delta, \partial_\gamma \nabla u_\delta) \Gamma_\delta^s \eta^2 \, dx \right.$$

$$\left. + \varepsilon^{-1} \int_\Omega D^2 f_\delta(\nabla u_\delta)(\partial_\gamma u_\delta \otimes \nabla \eta, \partial_\gamma u_\delta \otimes \nabla \eta) \Gamma_\delta^s \, dx \right\} \, .$$

If ε is sufficiently small, then we may absorb the first integral on the right-hand side and $i)$ is proved.

ad $ii)$. This time we choose $\varphi = \eta^2 \, \partial_\gamma u_\delta \max \left[\Gamma_\delta - k, 0 \right]$. The arguments given in $i)$ again show that this provides an admissible choice for (15):

$$\int_{A(k,r)} D^2 f_\delta(\nabla u_\delta)(\partial_\gamma \nabla u_\delta, \partial_\gamma \nabla u_\delta)(\Gamma_\delta - k) \eta^2 \, dx$$

$$+ \int_{A(k,r)} D^2 f_\delta(\nabla u_\delta)(\partial_\gamma \nabla u_\delta, \partial_\gamma u_\delta \otimes \nabla \Gamma_\delta) \eta^2 \, dx \tag{18}$$

$$= -2 \int_{A(k,r)} D^2 f_\delta(\nabla u_\delta)(\partial_\gamma \nabla u_\delta, \partial_\gamma u_\delta \otimes \nabla \eta) \eta (\Gamma_\delta - k) \, dx \, .$$

Here the non-negative first integral on the left-hand side is neglected and the second integral is estimated as above:

$$\frac{1}{2} \int_{A(k,r)} D^2 f_\delta(\nabla u_\delta)(e_j \otimes \nabla \Gamma_\delta, e_j \otimes \nabla \Gamma_\delta) \eta^2 \, dx$$

$$\leq \int_{A(k,r)} D^2 f_\delta(\nabla u_\delta)(\partial_\gamma \nabla u_\delta, \partial_\gamma u_\delta \otimes \nabla \Gamma_\delta) \eta^2 \, dx \, . \tag{19}$$

According to (17), the right-hand side of (18) satisfies almost everywhere

$$D^2 f_\delta(\nabla u_\delta)(\partial_\gamma \nabla u_\delta, \partial_\gamma u_\delta \otimes \nabla \eta) = \frac{1}{2} D^2 f_\delta(\nabla u_\delta)(e_j \otimes \nabla \eta, e_j \otimes \nabla \Gamma_\delta) \, . \tag{20}$$

Inequalities (18)–(20) imply with the Cauchy-Schwarz inequality

$$\int_{A(k,r)} D^2 f_\delta(\nabla u_\delta)(e_j \otimes \nabla \Gamma_\delta, e_j \otimes \nabla \Gamma_\delta)\, \eta^2 \, dx$$

$$\leq c \left\{ \varepsilon \int_{A(k,r)} D^2 f_\delta(\nabla u_\delta)(e_j \otimes \nabla \Gamma_\delta, e_j \otimes \nabla \Gamma_\delta)\, \eta^2 \, dx \right.$$

$$\left. + \varepsilon^{-1} \int_{A(k,r)} D^2 f_\delta(\nabla u_\delta)(e_j \otimes \nabla \eta, e_j \otimes \nabla \eta)\, (\Gamma_\delta - k)^2 \, dx \right\},$$

hence $ii)$ is proved by recalling Assumption 4.1, $iii)$, and by choosing $\varepsilon > 0$ sufficiently small. $\square$

4.2.2 The limit case $\mu = 3$

4.2.2.1 Higher local integrability

Now we concentrate on the limit case $\mu = 3$, prove local uniform integrability of $|\nabla u_\delta| \ln^2(1 + |\nabla u_\delta|^2)$ and establish the first two parts of Theorem 4.14 for any weak cluster point u^* of $\{u_\delta\}$. The last assertion then is a corollary of Theorem A.11, hence, we have a *Proof of Theorem 4.14.*

Note that the discussion of the vectorial setting does not depend on additional conditions (compare Remark 4.15).

Theorem 4.21. *Let Assumption 4.1 and Assumption 4.11 hold in the limit case $\mu = 3$. Then for any $\Omega' \Subset \Omega$ there is a real number $c(\Omega')$ – independent of δ – such that*

$$\int_{\Omega'} |\nabla u_\delta| \ln^2 \left(1 + |\nabla u_\delta|^2 \right) \, dx \leq c(\Omega') < \infty .$$

Proof. Given $B_{2r}(x_0) \Subset \Omega$ we first have to show that $\varphi = u_\delta \omega_\delta^2 \eta^2$, $\omega_\delta = \ln(\Gamma_\delta)$, $\eta \in C_0^\infty(B_{2r}(x_0))$, $0 \leq \eta \leq 1$, $\eta \equiv 1$ on $B_r(x_0)$, is admissible in the Euler equation (14). Since the vectorial structure condition is dropped in this section, we refer to the discussion of asymptotically regular integrands given in [CE]; a generalization is proved in [GiaM], Theorem 5.1. As a result, u_δ is seen to be of class $W^2_{2,loc} \cap W^1_{\infty,loc}(\Omega; \mathbb{R}^N)$ which proves that φ is an admissible choice. Alternatively, we could replace ω_δ by a suitable truncation $\omega_{\delta,M}$ and prove Theorem 4.21 by passing to the limit $M \to \infty$.

With the above choice, the Euler equation reads as

$$\int_{B_{2r}(x_0)} \nabla f(\nabla u_\delta) : \nabla u_\delta \, \omega_\delta^2 \eta^2 \, dx + \delta \int_{B_{2r}(x_0)} |\nabla u_\delta|^2 \, \omega_\delta^2 \eta^2 \, dx$$

$$= - \int_{B_{2r}(x_0)} \nabla f(\nabla u_\delta) : u_\delta \otimes \left[\nabla \omega_\delta^2 \eta^2 + \nabla \eta^2 \, \omega_\delta^2 \right] dx \qquad (21)$$

$$- \delta \int_{B_{2r}(x_0)} \nabla u_\delta : u_\delta \otimes \left[\nabla \omega_\delta^2 \eta^2 + \nabla \eta^2 \, \omega_\delta^2 \right] dx .$$

Remark 4.2, i), proves that the left-hand side of (21) is greater than or equal to

$$\int_{B_{2r}(x_0)} \left[\nu_1 \Gamma_\delta^{\frac{1}{2}} \omega_\delta^2 \eta^2 - \nu_2 \omega_\delta^2 \eta^2 \right] dx + \delta \int_{B_{2r}(x_0)} |\nabla u_\delta|^2 \omega_\delta^2 \eta^2 \, dx \, . \tag{22}$$

Since $|\nabla f|$ and $|u_\delta|$ are bounded, we find an upper bound for the right-hand side of (21) (using Young's inequality with $\varepsilon > 0$ fixed)

$$
\begin{aligned}
\text{r.h.s} \le c &\int_{B_{2r}(x_0)} \eta^2 \left[\varepsilon \Gamma_\delta^{\frac{1}{2}} \omega_\delta^2 + \varepsilon^{-1} \Gamma_\delta^{-\frac{1}{2}} |\nabla \omega_\delta|^2 \right] dx \\
&+ c(r) \int_{B_{2r}(x_0)} \omega_\delta^2 \, dx \\
&+ c\delta \int_{B_{2r}(x_0)} \eta^2 \left[\varepsilon |\nabla u_\delta|^2 \omega_\delta^2 + \varepsilon^{-1} |\nabla \omega_\delta|^2 \right] dx \\
&+ c(r) \delta \int_{B_{2r}(x_0)} |\nabla u_\delta| \omega_\delta^2 \, dx \, .
\end{aligned}
\tag{23}
$$

Clearly $\int_{B_{2r}(x_0)} \omega_\delta^2 \, dx$ and $\delta \int_{B_{2r}(x_0)} |\nabla u_\delta| \omega_\delta^2 \, dx$ are uniformly bounded with respect to δ (compare (13)). Hence, (21)–(23) imply after absorbing terms (for ε sufficiently small)

$$
\begin{aligned}
\int_{B_r(x_0)} \Gamma_\delta^{\frac{1}{2}} \omega_\delta^2 \, dx \le c \Bigg[1 &+ \int_{B_{2r}(x_0)} \Gamma_\delta^{-\frac{1}{2}} |\nabla \omega_\delta|^2 \eta^2 \, dx \\
&+ \delta \int_{B_{2r}(x_0)} |\nabla \omega_\delta|^2 \eta^2 \, dx \Bigg] \, .
\end{aligned}
\tag{24}
$$

Given (24) we observe that almost everywhere

$$|\nabla \omega_\delta|^2 \le c \frac{1}{1 + |\nabla u_\delta|^2} |\nabla^2 u_\delta|^2 \, ,$$

thus we may use Assumption 4.1, iii), with $\mu = 3$, Lemma 4.19 (letting $s = 0$ and recalling Remark 4.20) as well as Remark 4.2, iii), and (13) to obtain the final result

$$
\begin{aligned}
\int_{B_r(x_0)} \Gamma_\delta^{\frac{1}{2}} \omega_\delta^2 \, dx &\le c \left[1 + c \int_{B_{2r}(x_0)} \left(\Gamma_\delta^{-\frac{1}{2}} + \delta \right) \Gamma_\delta^{-1} |\nabla^2 u_\delta|^2 \eta^2 \, dx \right] \\
&\le c \left[1 + c \int_{B_{2r}(x_0)} D^2 f_\delta(\nabla u_\delta) (\partial_\gamma \nabla u_\delta, \partial_\gamma \nabla u_\delta) \eta^2 \, dx \right] \\
&\le c \left[1 + c(r) \int_{B_{2r}(x_0)} |D^2 f_\delta(\nabla u_\delta)| |\nabla u_\delta|^2 \, dx \right] \le c \, .
\end{aligned}
$$

$\square$

4.2.2.2 The independent variable

Throughout our whole studies we consider autonomous energy densities f, the case $f = f(x,P)$, $x \in \Omega$, $P \in \mathbb{R}^{nN}$ is omitted for the sake of technical simplicity. Here we take variational problems with linear growth and with limit ellipticity as a model to show that a smooth x-dependence does not affect our results.

Of course, a sufficiently smooth dependence on x can be considered in the superlinear case as well. We have chosen the linear setting with very weak ellipticity (i.e. $\mu = 3$) as a model which is closely related to the counterexample of Section 4.4.

For the moment we replace Assumption 4.1 by

Assumption 4.22. *There are constants $c_1, \ldots, c_7$ such that for all $x \in \overline{\Omega}$, for all P, U, $V \in \mathbb{R}^{nN}$ and for $\gamma = 1, \ldots, n$*

    i) the variational integrand $f = f(x,P)$ is of class $C^2\left(\overline{\Omega} \times \mathbb{R}^{nN}\right)$ and any of the derivatives occurring below exists;

    ii) $\left|\nabla_P f(x,P)\right| \leq c_1$;

    iii) $c_2 \left(1+|P|^2\right)^{-\frac{3}{2}} |U|^2 \leq D_P^2 f(x,P)(U,U) \leq c_3 \left(1+|P|^2\right)^{-\frac{1}{2}} |U|^2$;

    iv) $\left|\partial_\gamma \nabla_P f(x,P)\right| \leq c_4$;

    v) $\left|\partial_\gamma \partial_\gamma \nabla_P f(x,P)\right| \leq c_5$;

    vi) $\left|\partial_\gamma D_P^2 f(x,P)(U,V)\right| \leq c_6 \left|D_P^2 f(x,P)(U,V)\right| + \dfrac{c_7}{1+|P|^2} |U||V|$.

Moreover, we assume that the variational integrand $f = f(x,P)$ is of linear growth in P, uniformly with respect to x, i.e.

$$a|P| - b \leq f(x,P) \leq A|P| + B$$

holds with constants which are not depending on x.

Remark 4.23. *Maybe, assumption vi) needs some brief comment: if we want to include integrands of the type $f(x,P) = g(\alpha(x)P)$ with some scalar function α in our considerations, then we cannot expect that $\partial_\gamma D_P^2 f$ and $D_P^2 f$ define equivalent bilinear forms on $\mathbb{R}^{nN}$. However, the admissible perturbation on the right-hand side of vi) in particular gives a suitable approach to our example of Section 4.4 (compare Remark 4.43).*

We now claim

Theorem 4.24. *Theorem 4.14 remains true if Assumption 4.1 is replaced by Assumption 4.22.*

Proof. With the regularization introduced in Section 4.2.1 we first note that the arguments from duality theory remain unchanged (see, for instance, [ET]) and that we again have Remark 4.18. Moreover, Theorem A.11 is not affected by an additional x-dependence, hence we merely have to establish Theorem 4.21 with Assumption 4.22. To this purpose we recall the Euler equation in its non-differentiated form

$$\int_\Omega \nabla_P f_\delta(x, \nabla u_\delta) : \nabla \varphi \, dx = 0 \quad \text{for all } \varphi \in C_0^\infty(\Omega; \mathbb{R}^N) , \qquad (25)$$

which is of the same type as before. Now, the estimates (21)–(24) only depend on (25) if we take the following observation into account.

To prove Remark 4.2, *i*), we now replace $f(x, P)$ by $f_{x_0} := f(x, P) - \nabla_P f(x_0, 0) : P$. Note that the constants c_2–c_7 occurring in Assumption 4.22 remain unchanged. Moreover, Assumption 4.22, *ii*), holds uniformly with respect to x_0 for any f_{x_0} as above. As a consequence, we obtain

$$\nabla_P f(x_0, P) : P \geq c|P| \int_0^{|P|} (1 + \rho^2)^{-\frac{3}{2}} \, d\rho .$$

Then we choose r sufficiently small such that we have Remark 4.2, *i*), on $B_r(x_0)$.

Summarizing these results, the theorem is proved if we can show that

$$\int_{B_{2r}(x_0)} D_P^2 f_\delta(x, \nabla u_\delta)(\partial_\gamma \nabla u_\delta, \partial_\gamma \nabla u_\delta) \eta^2 \, dx \leq c(\eta) \qquad (26)$$

for any $B_{2r}(x_0) \Subset \Omega$ and all $\eta \in C_0^\infty(B_{2r}(x_0))$, $0 \leq \eta \leq 1$. At this point the differentiated form of (25) is needed which reads as

$$\int_\Omega D_P^2 f_\delta(x, \nabla u_\delta)(\partial_\gamma \nabla u_\delta, \nabla \varphi) \, dx + \int_\Omega (\partial_\gamma \nabla_P f_\delta)(x, \nabla u_\delta) : \nabla \varphi \, dx = 0 \quad (27)$$

for all $\varphi \in C_0^\infty(\Omega; \mathbb{R}^N)$. Once more the starting integrability of u_δ is good enough to take $\varphi = \eta^2 \partial_\gamma u_\delta$ (η as above) as an admissible test-function in (27). As a result we obtain

$$\int_{B_{2r}(x_0)} D_P^2 f_\delta(x, \nabla u_\delta)(\partial_\gamma \nabla u_\delta, \partial_\gamma \nabla u_\delta) \eta^2 \, dx$$

$$= -2 \int_{B_{2r}(x_0)} D_P^2 f_\delta(x, \nabla u_\delta)(\partial_\gamma \nabla u_\delta, \partial_\gamma u_\delta \otimes \nabla \eta) \eta \, dx$$

$$-2 \int_{B_{2r}(x_0)} (\partial_\gamma \nabla_P f_\delta)(x, \nabla u_\delta) : \partial_\gamma u_\delta \otimes \nabla \eta \, \eta \, dx \qquad (28)$$

$$- \int_{B_{2r}(x_0)} (\partial_\gamma \nabla_P f_\delta)(x, \nabla u_\delta) : \partial_\gamma \nabla u_\delta \, \eta^2 \, dx$$

$$=: I + II + III .$$

By Assumption 4.22, $iv)$, we have

$$|II| \leq c(\eta) \int_{B_{2r}(x_0)} |\nabla u_\delta| \, dx \leq c \, .$$

The first integral on the right-hand side of (28) is handled with Young's inequality for $\varepsilon > 0$ sufficiently small

$$|I| \leq \varepsilon \int_{B_{2r}(x_0)} D_P^2 f_\delta(x, \nabla u_\delta)(\partial_\gamma \nabla u_\delta, \partial_\gamma \nabla u_\delta) \eta^2 \, dx$$
$$+ c\varepsilon^{-1} \int_{B_{2r}(x_0)} |D_P^2 f_\delta(x, \nabla u_\delta)| \, |\nabla \eta|^2 \, |\nabla u_\delta|^2 \, dx \, .$$

Here the second integral on the right-hand side is uniformly bounded, the first one can be absorbed on the left-hand side of (28), hence it remains to find an upper bound for III. We perform an integration by parts to obtain

$$III = \int_{B_{2r}(x_0)} (\partial_\gamma \partial_\gamma \nabla_P f_\delta)(x, \nabla u_\delta) : \nabla u_\delta \, \eta^2 \, dx$$
$$+ \int_{B_{2r}(x_0)} (\partial_\gamma D_P^2 f_\delta)(x, \nabla u_\delta)(\partial_\gamma \nabla u_\delta, \nabla u_\delta) \eta^2 \, dx$$
$$+ \int_{B_{2r}(x_0)} (\partial_\gamma \nabla_P f_\delta)(x, \nabla u_\delta) : \nabla u_\delta \, \partial_\gamma \eta^2 \, dx$$

$$=: III_1 + III_2 + III_3 \, .$$

Assumption 4.22, $v)$, shows that $|III_1|$ is bounded independent of δ, the uniform estimate for $|III_3|$ again follows from $iv)$ of Assumption 4.22. Finally, for the consideration of $|III_2|$ we make use of Assumption 4.22, $vi)$, which together with Young's inequality gives for $\varepsilon > 0$ (note that the γ-derivative of the δ-part vanishes)

$$|III_2| \leq c \int_{B_{2r}(x_0)} |D_P^2 f(x, \nabla u_\delta)(\partial_\gamma \nabla u_\delta, \nabla u_\delta)| \eta^2 \, dx$$
$$+ c \int_{B_{2r}(x_0)} \left(1 + |\nabla u_\delta|^2\right)^{-1} |\nabla^2 u_\delta| \, |\nabla u_\delta| \eta^2 \, dx$$
$$\leq c\varepsilon \int_{B_{2r}(x_0)} D_P^2 f(x, \nabla u_\delta)(\partial_\gamma \nabla u_\delta, \partial_\gamma \nabla u_\delta) \eta^2 \, dx$$
$$+ c\varepsilon^{-1} \int_{B_{2r}(x_0)} D_P^2 f(x, \nabla u_\delta)(\nabla u_\delta, \nabla u_\delta) \eta^2 \, dx \tag{29}$$
$$+ c\varepsilon \int_{B_{2r}(x_0)} \left(1 + |\nabla u_\delta|^2\right)^{-\frac{3}{2}} |\nabla^2 u_\delta|^2 \eta^2 \, dx$$
$$+ c\varepsilon^{-1} \int_{B_{2r}(x_0)} \left(1 + |\nabla u_\delta|^2\right)^{-\frac{1}{2}} |\nabla u_\delta|^2 \eta^2 \, dx \, .$$

Here, on account of the ellipticity Assumption 4.22, *iii*), the third integral on the right-hand side is estimated by the first one which in return is absorbed on the left-hand side of (28). The remaining two integrals on the right-hand side of (29) are handled with the linear growth of f which, by assumption, is uniformly with respect to x (we also recall the second inequality of Assumption 4.22, *iii*)). Thus, the theorem is proved. □

As mentioned above, we will come back to the x-dependent situation in Section 4.4.

4.2.3 L^p-estimates in the case $\mu < 3$

From now on we assume a slightly stronger ellipticity condition, i.e. we consider the case $\mu < 3$. Moreover, we again have the additional structure in the vectorial setting $N > 1$. Then it is possible to modify the arguments of Section 4.2.2.1 such that the results obtained there may be iterated. This gives uniform L^p-estimates in the sense of

Theorem 4.25. *Suppose that $\mu < 3$, that we have Assumption 4.1, Assumption 4.11 and that (2), Section 4.1, is satisfied. Then for any $1 < p < \infty$ and for any $\Omega' \Subset \Omega$ there is a constant $c(p, \Omega')$, which does not depend on δ, such that*

$$\|\nabla u_\delta\|_{L^p(\Omega'; \mathbb{R}^{nN})} \leq c(p, \Omega') < \infty .$$

Remark 4.26. *As an immediate consequence we can find a generalized minimizer $u^* \in \mathcal{M}$ which is of class $W^1_{p,loc}(\Omega; \mathbb{R}^N)$ for any $1 < p < \infty$.*

Proof of Theorem 4.25. Fix a ball $B_{r_0}(x_0) \Subset \Omega$ and assume that there is a real number $\alpha_0 \geq 0$ such that (uniformly with respect to δ)

$$\int_{B_{r_0}(x_0)} \Gamma_\delta^{\frac{1+\alpha_0}{2}} \, dx + \delta \int_{B_{r_0}(x_0)} \Gamma_\delta^{1+\frac{\alpha_0}{2}} \, dx \leq c . \qquad (30)$$

Note that by (13) this assumption is true for $\alpha_0 = 0$. Next define $\alpha = \alpha_0 + 3 - \mu$ and choose $\varphi = u_\delta \, \Gamma_\delta^{\frac{\alpha}{2}} \eta^2$, $\eta \in C_0^\infty(B_{r_0}(x_0))$, $0 \leq \eta \leq 1$, $\eta \equiv 1$ on $B_{r_0/2}(x_0)$, $|\nabla \eta| \leq c/r_0$. As outlined above, u_δ is of class $W^2_{2,loc} \cap W^1_{\infty,loc}(\Omega; \mathbb{R}^N)$, hence φ is admissible in (14) with the result

$$\int_{B_{r_0}(x_0)} \nabla f(\nabla u_\delta) : \nabla u_\delta \, \Gamma_\delta^{\frac{\alpha}{2}} \eta^2 \, dx + \delta \int_{B_{r_0}(x_0)} |\nabla u_\delta|^2 \, \Gamma_\delta^{\frac{\alpha}{2}} \eta^2 \, dx$$

$$\leq c(\alpha) \int_{B_{r_0}(x_0)} \Gamma_\delta^{\frac{\alpha-1}{2}} |\nabla^2 u_\delta| \eta^2 \, dx + c(\alpha) \delta \int_{B_{r_0}(x_0)} \Gamma_\delta^{\frac{\alpha}{2}} |\nabla^2 u_\delta| \eta^2 \, dx \qquad (31)$$

$$+ c \int_{B_{r_0}(x_0)} \Gamma_\delta^{\frac{\alpha}{2}} |\nabla \eta^2| \, dx + c\delta \int_{B_{r_0}(x_0)} \Gamma_\delta^{\frac{\alpha+1}{2}} |\nabla \eta^2| \, dx .$$

Here Assumption 4.11 and the boundedness of ∇f (compare Assumption 4.1, *ii*)) are used. Analogous to Section 4.2.2.1, the left-hand side of (31) is estimated with the help of Remark 4.2, *i*):

$$l.h.s. \geq \nu_1 \int_{B_{r_0}(x_0)} \Gamma_\delta^{\frac{1+\alpha}{2}} \eta^2 \, dx - \nu_2 \int_{B_{r_0}(x_0)} \Gamma_\delta^{\frac{\alpha}{2}} \eta^2 \, dx$$

$$+ \delta \int_{B_{r_0}(x_0)} \Gamma_\delta^{1+\frac{\alpha}{2}} \eta^2 \, dx - \delta \int_{B_{r_0}(x_0)} \Gamma_\delta^{\frac{\alpha}{2}} \eta^2 \, dx \, .$$

The right-hand side of (31) is handled via (fix $\varepsilon > 0$ and use Young's inequality)

$$r.h.s \leq c \int_{B_{r_0}(x_0)} \eta^2 \left[\varepsilon \Gamma_\delta^{\frac{1+\alpha}{2}} + \varepsilon^{-1} \Gamma_\delta^{-\frac{1+\alpha}{2}} \Gamma_\delta^{\alpha-1} |\nabla^2 u_\delta|^2 \right] \, dx$$

$$+ c \int_{B_{r_0}(x_0)} \left[\varepsilon \Gamma_\delta^{\frac{1+\alpha}{2}} \eta^2 + \varepsilon^{-1} \Gamma_\delta^{-\frac{1+\alpha}{2}} \Gamma_\delta^{\alpha} |\nabla \eta|^2 \right] \, dx$$

$$+ c\delta \int_{B_{r_0}(x_0)} \eta^2 \left[\varepsilon \Gamma_\delta^{1+\frac{\alpha}{2}} + \varepsilon^{-1} \Gamma_\delta^{-1-\frac{\alpha}{2}} \Gamma_\delta^{\alpha} |\nabla^2 u_\delta|^2 \right] \, dx$$

$$+ c\delta \int_{B_{r_0}(x_0)} \left[\varepsilon \Gamma_\delta^{1+\frac{\alpha}{2}} \eta^2 + \varepsilon^{-1} \Gamma_\delta^{-1-\frac{\alpha}{2}} \Gamma_\delta^{1+\alpha} |\nabla \eta|^2 \right] \, dx \, .$$

Hence, absorbing terms, (31) yields

$$\int_{B_{r_0/2}(x_0)} \Gamma_\delta^{\frac{1+\alpha}{2}} \, dx + \delta \int_{B_{r_0/2}(x_0)} \Gamma_\delta^{1+\frac{\alpha}{2}} \, dx$$

$$\leq c \left[\int_{B_{r_0}(x_0)} \eta^2 \Gamma_\delta^{\frac{\alpha-3}{2}} |\nabla^2 u_\delta|^2 \, dx \right.$$

$$+ \int_{B_{r_0}(x_0)} \Gamma_\delta^{\frac{\alpha-1}{2}} |\nabla \eta|^2 \, dx + \left. \int_{B_{r_0}(x_0)} \Gamma_\delta^{\frac{\alpha}{2}} \eta^2 \, dx \right]$$

$$+ c\delta \left[\int_{B_{r_0}(x_0)} \eta^2 \Gamma_\delta^{\frac{\alpha}{2}-1} |\nabla^2 u_\delta|^2 \, dx \right. \tag{32}$$

$$+ \int_{B_{r_0}(x_0)} \Gamma_\delta^{\frac{\alpha}{2}} |\nabla \eta|^2 \, dx + \left. \int_{B_{r_0}(x_0)} \Gamma_\delta^{\frac{\alpha}{2}} \eta^2 \, dx \right]$$

$$=: c \sum_{i=1}^{3} I_i + c\delta \sum_{i=4}^{6} I_i \, .$$

Starting with I_1, we recall that by definition $\mu + \alpha - 3 = \alpha_0 \geq 0$, thus Assumption 4.1, $iii)$, and Lemma 4.19, $i)$, give

$$I_1 = \int_{B_{r_0}(x_0)} \eta^2 \Gamma_\delta^{-\frac{\mu}{2}} |\nabla^2 u_\delta|^2 \Gamma_\delta^{\frac{\mu+\alpha-3}{2}} \, dx$$

$$\leq c \int_{B_{r_0}(x_0)} D^2 f_\delta(\nabla u_\delta) (\partial_\gamma \nabla u_\delta, \partial_\gamma \nabla u_\delta) \Gamma_\delta^{\frac{\alpha_0}{2}} \eta^2 \, dx$$

$$\leq c(r_0) \int_{B_{r_0}(x_0)} \left[\Gamma_\delta^{-\frac{1}{2}} + \delta \right] \Gamma_\delta^{1+\frac{\alpha_0}{2}} \, dx \leq c \, ,$$

where the last inequality is due to the assumption (30). An upper bound (not depending on δ) for I_3 is found since we may assume without loss of generality that $\mu \geq 2$. This clearly proves I_2 to be bounded independent of δ as well. Studying I_4 let us first assume that $\alpha \leq 2$. Then, again by Lemma 4.19, i),

$$\delta I_4 \leq \int_{B_{r_0}(x_0)} D^2 f_\delta(\nabla u_\delta)(\partial_\gamma \nabla u_\delta, \partial_\gamma \nabla u_\delta)\,\eta^2\,dx$$

$$\leq c(r_0) \int_{B_{r_0}(x_0)} \left[\Gamma_\delta^{-\frac{1}{2}} + \delta\right] \Gamma_\delta\,dx \leq c\,.$$

In the case $\alpha > 2$, Lemma 4.19, i), gives

$$\delta I_4 \leq \int_{B_{r_0}(x_0)} D^2 f_\delta(\nabla u_\delta)(\partial_\gamma \nabla u_\delta, \partial_\gamma \nabla u_\delta)\,\Gamma_\delta^{\frac{\alpha}{2}-1}\eta^2\,dx$$

$$\leq c(r_0) \int_{B_{r_0}(x_0)} \left[\Gamma_\delta^{-\frac{1}{2}} + \delta\right] \Gamma_\delta^{1+\frac{\alpha}{2}-1}\,dx \leq c\,,$$

where we once more recall (30) and observe that $\frac{\alpha}{2}-1 = (\alpha_0+1-\mu)/2 \leq \alpha_0/2$. This condition trivially bounds δI_5 and δI_6 (independent of δ) and we have proved with (32): suppose that (30) holds for some given $r_0 > 0$ and $\alpha_0 \geq 0$. Then there is a constant, independent of δ, such that

$$\int_{B_{r_0/2}(x_0)} \Gamma_\delta^{\frac{1+\alpha_0+3-\mu}{2}}\,dx + \delta \int_{B_{r_0/2}(x_0)} \Gamma_\delta^{1+\frac{\alpha_0+3-\mu}{2}}\,dx \leq c\,. \qquad (33)$$

We now claim that for any $m \in \mathbb{N}$ there is a constant $c(m)$, independent of δ, such that

$$\int_{B_{r_0/2^m}(x_0)} \Gamma_\delta^{\frac{1+m(3-\mu)}{2}}\,dx + \delta \int_{B_{r_0/2^m}(x_0)} \Gamma_\delta^{1+\frac{m(3-\mu)}{2}}\,dx \leq c\,. \qquad (34)$$

In fact, as mentioned above, $\alpha_0 = 0$ is an admissible choice to obtain (34) from (33) in the case $m = 1$. Next assume by induction that (34) is true for some $m \in \mathbb{N}$. Then we may take $\alpha_0 = m(3 - \mu)$ in (30) and (33) gives (34) with m replaced by $m + 1$, thus the claim is proved. Obviously, this implies Theorem 4.25. $\square$

Remark 4.27. *If we omit the structure condition in the vectorial setting, then analogous arguments prove higher integrability up to a finite number $1 < p(\mu)$.*

4.2.4 A priori gradient bounds

Here the DeGiorgi-type arguments as outlined in Section 3.3.3 are modified: on one hand, given Theorem 4.25, we benefit from Hölder's inequality. This decreases on the other hand the exponent β of iteration (see the definition of β given below). Nevertheless, it turns out that Lemma 3.26 still is applicable to obtain

Theorem 4.28. *Consider a ball $B_{R_0}(x_0) \Subset \Omega$. With the assumptions of Theorem 4.25 there is a local constant $c > 0$ such that for any $\delta \in (0,1)$*

$$\|\nabla u_\delta\|_{L^\infty(B_{R_0/2}, \mathbb{R}^{nN})} \leq c.$$

Before proving Theorem 4.28 we recall the definitions $\Gamma_\delta = 1 + |\nabla u_\delta|^2$,

$$A(k,r) = \left\{ x \in B_r(x_0) : \Gamma_\delta > k \right\}, \quad B_r(x_0) \Subset \Omega, \quad k > 0,$$

and establish the following result.

Lemma 4.29. *Fix some $x_0 \in \Omega$ and suppose that we are given radii $0 < r < R < R_0$, $B_{R_0}(x_0) \Subset \Omega$. Then there is a real number c, independent of r, R, R_0, k and δ, such that*

$$\int_{A(k,r)} (\Gamma_\delta - k)^{\frac{n}{n-1}} \, dx$$

$$\leq \frac{c}{(R-r)^{\frac{n}{n-1}}} \left[\int_{A(k,R)} (\Gamma_\delta - k)^2 \, dx \right]^{\frac{1}{2}\frac{n}{n-1}} \left[\int_{A(k,R)} \Gamma_\delta^{\frac{\mu}{2}} \, dx \right]^{\frac{1}{2}\frac{n}{n-1}}. \tag{35}$$

Proof of Lemma 4.29. Recalling the notion $w^+ = \max[w, 0]$, Sobolev's inequality yields for all $\eta \in C_0^\infty(B_R(x_0))$ such that $0 \leq \eta \leq 1$, $\eta \equiv 1$ on $B_r(x_0)$, $|\nabla \eta| \leq c/(R-r)$,

$$\int_{A(k,r)} (\Gamma_\delta - k)^{\frac{n}{n-1}} \, dx \leq \int_{B_R(x_0)} \left[\eta \, (\Gamma_\delta - k)^+ \right]^{\frac{n}{n-1}} \, dx$$

$$\leq c \left[\int_{B_R(x_0)} \left| \nabla \left[\eta \, (\Gamma_\delta - k)^+ \right] \right| \, dx \right]^{\frac{n}{n-1}}$$

$$\leq c \left[\int_{A(k,R)} \left| \nabla \left[\eta \, (\Gamma_\delta - k) \right] \right| \, dx \right]^{\frac{n}{n-1}} \tag{36}$$

$$\leq c \left[I_1^{\frac{n}{n-1}} + I_2^{\frac{n}{n-1}} \right].$$

Here we have

$$I_1^{\frac{n}{n-1}} := \left[\int_{A(k,R)} |\nabla \eta| \, (\Gamma_\delta - k) \, dx \right]^{\frac{n}{n-1}}$$

$$\leq \left[\int_{A(k,R)} |\nabla \eta|^2 \, (\Gamma_\delta - k)^2 \, dx \right]^{\frac{1}{2}\frac{n}{n-1}} \left[\int_{A(k,R)} 1 \, dx \right]^{\frac{1}{2}\frac{n}{n-1}}$$

$$\leq \frac{c}{(R-r)^{\frac{n}{n-1}}} \left[\int_{A(k,R)} (\Gamma_\delta - k)^2 \, dx \right]^{\frac{1}{2}\frac{n}{n-1}} \left[\int_{A(k,R)} 1 \, dx \right]^{\frac{1}{2}\frac{n}{n-1}},$$

thus $I_1^{\frac{n}{n-1}}$ is bounded from above by the right-hand side of (35). Estimating I_2, Lemma 4.19, *ii*), is needed with the result

$$I_2^{\frac{n}{n-1}} := \left[\int_{A(k,R)} \eta |\nabla \Gamma_\delta| \, dx \right]^{\frac{n}{n-1}}$$

$$\leq \left[\int_{A(k,R)} \eta^2 |\nabla \Gamma_\delta|^2 \Gamma_\delta^{-\frac{\mu}{2}} \, dx \right]^{\frac{1}{2}\frac{n}{n-1}} \left[\int_{A(k,R)} \Gamma_\delta^{\frac{\mu}{2}} \, dx \right]^{\frac{1}{2}\frac{n}{n-1}}$$

$$\leq c \left[\int_{A(k,R)} D^2 f_\delta(\nabla u_\delta)(e_j \otimes \nabla \eta, e_j \otimes \nabla \eta)(\Gamma_\delta - k)^2 \, dx \right]^{\frac{1}{2}\frac{n}{n-1}}$$

$$\cdot \left[\int_{A(k,R)} \Gamma_\delta^{\frac{\mu}{2}} \, dx \right]^{\frac{1}{2}\frac{n}{n-1}}$$

$$\leq \frac{c}{(R-r)^{\frac{n}{n-1}}} \left[\int_{A(k,R)} (\Gamma_\delta - k)^2 \, dx \right]^{\frac{1}{2}\frac{n}{n-1}} \left[\int_{A(k,R)} \Gamma_\delta^{\frac{\mu}{2}} \, dx \right]^{\frac{1}{2}\frac{n}{n-1}},$$

hence (36) proves the lemma. $\square$

We now come to the *Proof of Theorem 4.28*. Consider the left-hand side of (35): for any real number $s > 1$, Hölder's inequality implies

$$\int_{A(k,r)} (\Gamma_\delta - k)^2 \, dx = \int_{A(k,r)} (\Gamma_\delta - k)^{\frac{n}{n-1}\frac{1}{s}} (\Gamma_\delta - k)^{2 - \frac{n}{n-1}\frac{1}{s}} \, dx$$

$$\leq \left[\int_{A(k,r)} (\Gamma_\delta - k)^{\frac{n}{n-1}} \, dx \right]^{\frac{1}{s}}$$

$$\cdot \left[\int_{A(k,r)} (\Gamma_\delta - k)^{\left(2 - \frac{n}{n-1}\frac{1}{s}\right)\frac{s}{s-1}} \right]^{\frac{s-1}{s}}.$$

Thus, on account of Theorem 4.25, there is a real number $c_1(s, n, B_{R_0}(x_0))$, independent of δ,

$$c_1(s, n, B_{R_0}(x_0)) := \sup_{\delta > 0} \left[\int_{B_{R_0}(x_0)} \Gamma_\delta^{\frac{1}{s-1}\left(2s - \frac{n}{n-1}\right)} \, dx \right]^{\frac{s-1}{s}} < \infty,$$

such that

$$\int_{A(k,r)} (\Gamma_\delta - k)^2 \, dx \leq c_1(s, n, B_{R_0}(x_0)) \left[\int_{A(k,r)} (\Gamma_\delta - k)^{\frac{n}{n-1}} \, dx \right]^{\frac{1}{s}}. \quad (37)$$

Studying the right-hand side of (35), we fix a second real number $t > 1$ and applying Hölder's inequality once more it is established that

$$\int_{A(k,R)} \Gamma_\delta^{\frac{\mu}{2}} \, dx \leq |A(k,R)|^{\frac{1}{t}} \left[\int_{A(k,R)} \Gamma_\delta^{\frac{\mu}{2}\frac{t}{t-1}} \, dx \right]^{\frac{t-1}{t}}.$$

Let

$$c_2(t, \mu, B_{R_0}(x_0)) := \sup_{\delta > 0} \left[\int_{B_{R_0}(x_0)} \Gamma_\delta^{\frac{\mu}{2} \frac{t}{t-1}} \, dx \right]^{\frac{t-1}{t}} < \infty \, .$$

Then we have

$$\int_{A(k,R)} \Gamma_\delta^{\frac{\mu}{2}} \, dx \le c_2(t, \mu, B_{R_0}(x_0)) \left| A(k,R) \right|^{\frac{1}{t}} \, . \tag{38}$$

Summarizing these results we arrive at

$$\int_{A(k,r)} \left(\Gamma_\delta - k \right)^2 \, dx \overset{(37)}{\le} c \left[\int_{A(k,r)} \left(\Gamma_\delta - k \right)^{\frac{n}{n-1}} \, dx \right]^{\frac{1}{s}}$$

$$\overset{(35)}{\le} \frac{c}{(R-r)^{\frac{n}{n-1} \frac{1}{s}}} \left[\int_{A(k,R)} \left(\Gamma_\delta - k \right)^2 \, dx \right]^{\frac{1}{2} \frac{n}{n-1} \frac{1}{s}}$$

$$\cdot \left[\int_{A(k,R)} \Gamma_\delta^{\frac{\mu}{2}} \, dx \right]^{\frac{1}{2} \frac{n}{n-1} \frac{1}{s}} \tag{39}$$

$$\overset{(38)}{\le} \frac{c}{(R-r)^{\frac{n}{n-1} \frac{1}{s}}} \left[\int_{A(k,R)} \left(\Gamma_\delta - k \right)^2 \, dx \right]^{\frac{1}{2} \frac{n}{n-1} \frac{1}{s}}$$

$$\cdot \left| A(k,R) \right|^{\frac{1}{2} \frac{n}{n-1} \frac{1}{s} \frac{1}{t}} \, .$$

As the next step we define for k and $r < R$ as above the following quantities:

$$\tau(k,r) := \int_{A(k,r)} \left(\Gamma_\delta - k \right)^2 \, dx \, ,$$

$$a(k,r) := \left| A(k,r) \right| \, .$$

With this notation, (39) can be written in the form

$$\tau(k,r) \le \frac{c}{(R-r)^{\frac{n}{n-1} \frac{1}{s}}} \left[\tau(k,R) \right]^{\frac{1}{2} \frac{n}{n-1} \frac{1}{s}} \left[a(k,R) \right]^{\frac{1}{2} \frac{n}{n-1} \frac{1}{s} \frac{1}{t}} \, . \tag{40}$$

Given two real numbers $h > k > 0$, we now observe

$$a(h,R) = \int_{A(h,R)} dx \le \int_{A(h,R)} \left(\Gamma_\delta - k \right)^2 (h-k)^{-2} \, dx \, ,$$

thus we get for $h > k > 0$

$$a(h,R) \le \frac{1}{(h-k)^2} \tau(k,R) \, . \tag{41}$$

With (40) and (41) it is proved that for $h > k > 0$ we have the estimate

$$\tau(h,r) \leq \frac{c}{(R-r)^{\frac{n}{n-1}\frac{1}{s}}} \left[\tau(h,R)\right]^{\frac{1}{2}\frac{n}{n-1}\frac{1}{s}} \frac{1}{(h-k)^{\frac{n}{n-1}\frac{1}{s}\frac{1}{t}}} \left[\tau(k,R)\right]^{\frac{1}{2}\frac{n}{n-1}\frac{1}{s}\frac{1}{t}}$$

$$\leq \frac{c}{(R-r)^{\frac{n}{n-1}\frac{1}{s}}} \frac{1}{(h-k)^{\frac{n}{n-1}\frac{1}{s}\frac{1}{t}}} \left[\tau(k,R)\right]^{\frac{1}{2}\frac{n}{n-1}\frac{1}{s}\left(1+\frac{1}{t}\right)} .$$

Now s and t are chosen sufficiently close to 1 (depending on n) such that

$$\frac{1}{2}\frac{n}{n-1}\frac{1}{s}\left[1+\frac{1}{t}\right] =: \beta > 1 .$$

With this choice of s and t we additionally let

$$\alpha := \frac{n}{n-1}\frac{1}{s}\frac{1}{t} > 0 , \qquad \gamma := \frac{n}{n-1}\frac{1}{s} > 0 .$$

Thus we are in the situation of Lemma 3.26. Applying this lemma we have

$$\tau(d, R_0/2) = \int_{A(d,R_0/2)} \left(\Gamma_\delta - d\right)^2 \, dx = 0 ,$$

and, as a consequence,

$$\Gamma_\delta \leq d \quad \text{on } B_{R_0/2}(x_0) . \tag{42}$$

Here the quantity d is uniformly bounded with respect to δ if and only if there is a constant (independent of δ) such that

$$\tau(0, R_0) = \int_{B_{R_0}(x_0)} \Gamma_\delta^2 \, dx \leq c .$$

This fact however was proved in Theorem 4.25 and the a priori estimate Theorem 4.28 follows from (42). $\square$

Remark 4.30. *With Theorem 4.28 we also have a complete* Proof of Theorem 4.16. *This was already discussed in Section 4.1.3. Note that Proposition 4.10 again can be replaced by the arguments given in Section 2.3.2.*

4.3 Two-dimensional problems

To finish this chapter we study two-dimensional μ-elliptic variational problems with linear growth. It turns out that (assuming the usual structure condition if $N > 1$) in this case no boundedness condition is necessary to find the unique (up to a constant) solution of the variational problem

$$\int_\Omega f(\nabla w) \, dx + \int_{\partial\Omega} f_\infty((u_0 - w) \otimes \nu) \, d\mathcal{H}^{n-1} \to \min \quad \text{in } W_1^1(\Omega; \mathbb{R}^N) \quad (\mathcal{P}')$$

in the limit case $\mu = 3$.

We then reduce the two-dimensional (unbounded) case to the setting of Section 4.2 and get, if $n = 2$, the validity of Theorem 4.16 ($\mu < 3$) in the most general setting. To be precise: from now on we consider boundary data $u_0 \in W_2^1(\Omega; \mathbb{R}^N)$, where again the case $u_0 \in W_1^1(\Omega; \mathbb{R}^N)$ is handled as outlined in Remark 2.5 (compare [Bi5]). We are going to prove

Theorem 4.31. *Suppose that Assumption 4.1 holds and that we have in addition (2), Section 4.1, in the vectorial case $N > 1$. Moreover, consider the limit case $\mu = 3$. Then, if $n = 2$, there is a generalized minimizer $u^* \in \mathcal{M}$ such that*

i) We have $\nabla^s u^* \equiv 0$, hence $\nabla u^* \equiv \nabla^a u^*$.

ii) For any $\Omega' \Subset \Omega$ there is a constant $c(\Omega')$ satisfying

$$\int_{\Omega'} |\nabla u^*| \ln\left(1 + |\nabla u^*|^2\right) \, \mathrm{d}x \leq c(\Omega') < \infty \, .$$

iii) The particular minimizer u^* is of class $W_1^1(\Omega; \mathbb{R}^N)$ and (up to a constant) the unique solution of the variational problem $(\mathcal{P}')$.

Theorem 4.32. *Suppose that Assumption 4.1 is valid with $\mu < 3$. In the case $N > 1$ we additionally assume (2) and (3) as stated in Section 4.1. Then, if $n = 2$, we have*

i) Each generalized minimizer $u \in \mathcal{M}$ is in the space $C^{1,\alpha}(\Omega; \mathbb{R}^N)$ for any $0 < \alpha < 1$.

ii) For $u, v \in \mathcal{M}$ we have $\nabla u = \nabla v$, i.e. generalized minimizers are unique up to a constant.

4.3.1 Higher local integrability in the limit case

Here we are going to establish uniform local higher integrability of the sequence $\{u_\delta\}$ (see Section 4.2.1) in the limit case $\mu = 3$.

Let us, for the moment, concentrate on the scalar case $N = 1$. Then we have the following assertion

Lemma 4.33. *Suppose that the assumptions of Theorem 4.31 hold in the two-dimensional scalar case $n = 2$, $N = 1$, and let $\{u_\delta\}$ denote the regularization introduced above. Moreover, fix a ball $B_r(x_0)$ satisfying $B_{2r}(x_0) \Subset \Omega$. Then there is a positive number $c = c(r)$, independent of δ, such that for all $\eta \in C_0^\infty(B_{2r}(x_0))$, $0 \leq \eta \leq 1$,*

$$\int_{B_{2r}(x_0)} \left(1 + |\nabla u_\delta|^2\right)^{\frac{1}{2}} |u_\delta - (u_\delta)_{2r}|^2 \eta^2 \, \mathrm{d}x$$

$$+ \delta \int_{B_{2r}(x_0)} |\nabla u_\delta|^2 |u_\delta - (u_\delta)_{2r}|^2 \eta^2 \, \mathrm{d}x \leq c \, .$$

Here $(u_\delta)_{2r}$ denotes the mean value of u_δ on $B_{2r}(x_0)$.

Remark 4.34.

i) *Following the proof of Theorem 4.36 below, it becomes obvious that this estimate is exactly the one which is needed to reach the limit case $\mu = 3$.*

ii) *Inequality (43) given below is the main reason why the results in two dimensions are better than the ones stated above for $n \geq 3$.*

Proof of Lemma 4.33. Note that in the two-dimensional case $n = 2$ we have by Sobolev-Poincarè's inequality

$$\left(\int_{B_{2r}(x_0)} |u_\delta - (u_\delta)_{2r}|^2 \, dx \right)^{\frac{1}{2}} \leq c_1 \int_{B_{2r}(x_0)} |\nabla u_\delta| \, dx \leq c_2 \qquad (43)$$

for some positive constants c_1, c_2 which are not depending on δ (recall Remark 4.2, ii), and (13), Section 4.2.1). Since u_δ is a solution of the approximative problem we have $u_\delta \in W_t^2(B_{2r}(x_0))$ for any $t < \infty$. This was outlined in Section 3.3.1. As a result, $\varphi = (u_\delta - (u_\delta)_{2r})^3 \eta^2$, $\eta \in C_0^\infty(B_{2r}(x_0))$, $0 \leq \eta \leq 1$, is admissible in the Euler equation (see (14), Section 4.2.1) and we obtain

$$3 \int_{B_{2r}(x_0)} \nabla f(\nabla u_\delta) \nabla u_\delta |u_\delta - (u_\delta)_{2r}|^2 \eta^2 \, dx$$

$$+ 3\delta \int_{B_{2r}(x_0)} |\nabla u_\delta|^2 |u_\delta - (u_\delta)_{2r}|^2 \eta^2 \, dx$$

$$= -2 \int_{B_{2r}(x_0)} \nabla f_\delta(\nabla u_\delta) \nabla \eta \, \eta \, (u_\delta - (u_\delta)_{2r})^3 \, dx \ .$$

Remark 4.2, i), (43) and the boundedness of $|\nabla f|$ imply

$$\int_{B_{2r}(x_0)} \left(1 + |\nabla u_\delta|^2 \right)^{\frac{1}{2}} |u_\delta - (u_\delta)_{2r}|^2 \eta^2 \, dx$$

$$+ \delta \int_{B_{2r}(x_0)} |\nabla u_\delta|^2 |u_\delta - (u_\delta)_{2r}|^2 \eta^2 \, dx \qquad (44)$$

$$\leq c(1 + I_1 + I_2) \ ,$$

where the constant c again is not depending on δ, and I_1, I_2 are given by

$$I_1 = \int_{B_{2r}(x_0)} |u_\delta - (u_\delta)_{2r}|^3 \eta |\nabla \eta| \, dx \ ,$$

$$I_2 = \delta \int_{B_{2r}(x_0)} |\nabla u_\delta| |u_\delta - (u_\delta)_{2r}|^3 \eta |\nabla \eta| \, dx \ .$$

Estimating I_1 we observe (using (43), Hölder's inequality, Sobolev-Poincarè's inequality and Young's inequality for some sufficiently small number $\varepsilon > 0$)

$$I_1 \leq \left(\int_{B_{2r}(x_0)} |u_\delta - (u_\delta)_{2r}|^4 \eta^2 \, dx \right)^{\frac{1}{2}}$$

$$\cdot \left(\int_{B_{2r}(x_0)} |u_\delta - (u_\delta)_{2r}|^2 |\nabla \eta|^2 \, dx \right)^{\frac{1}{2}}$$

$$\leq c \int_{B_{2r}(x_0)} \left| \nabla \left(|u_\delta - (u_\delta)_{2r}|^2 \eta \right) \right| \, dx \tag{45}$$

$$\leq c \left(1 + \int_{B_{2r}(x_0)} |u_\delta - (u_\delta)_{2r}| |\nabla u_\delta| \eta \, dx \right)$$

$$\leq c \left(1 + \int_{B_{2r}(x_0)} \left\{ \varepsilon |u_\delta - (u_\delta)_{2r}|^2 \left(1 + |\nabla u_\delta|^2 \right)^{\frac{1}{2}} \eta^2 \right. \right.$$

$$\left. \left. + \varepsilon^{-1} \left(1 + |\nabla u_\delta|^2 \right)^{\frac{1}{2}} \right\} \, dx \right) .$$

Here again c denotes some positive local constant which is not depending on δ. Note that the "ε"-part on the right-hand side of (45) can be absorbed (for ε sufficiently small) on the left-hand side of (44), whereas the remaining integral is uniformly bounded with respect to δ.

To find an estimate for I_2 we recall that

$$\delta \int_\Omega |\nabla u_\delta|^2 \, dx \to 0 \quad \text{as } \delta \to 0 .$$

This was proved in (9), Section 2.1.2. As a consequence, we have with suitable local constants and for $\varepsilon > 0$ sufficiently small

$$I_2 \leq c\delta \left(\int_{B_{2r}(x_0)} |\nabla u_\delta|^2 \, dx \right)^{\frac{1}{2}} \left(\int_{B_{2r}(x_0)} |u_\delta - (u_\delta)_{2r}|^6 \eta^2 \, dx \right)^{\frac{1}{2}}$$

$$\leq c\delta^{\frac{1}{2}} \int_{B_{2r}(x_0)} \left| \nabla \left(|u_\delta - (u_\delta)_{2r}|^3 \eta \right) \right| \, dx$$

$$\leq c\delta^{\frac{1}{2}} \left(\int_{B_{2r}(x_0)} |u_\delta - (u_\delta)_{2r}|^2 |\nabla u_\delta| \eta \, dx + \int_{B_{2r}(x_0)} |u_\delta - (u_\delta)_{2r}|^3 \, dx \right)$$

$$\leq c\delta^{\frac{1}{2}} \left(\int_{B_{2r}(x_0)} \left(\varepsilon \delta^{\frac{1}{2}} |u_\delta - (u_\delta)_{2r}|^2 |\nabla u_\delta|^2 \eta^2 \right. \right.$$

$$\left. \left. + \varepsilon^{-1} \delta^{-\frac{1}{2}} |u_\delta - (u_\delta)_{2r}|^2 \right) \, dx + \int_{B_{2r}(x_0)} |u_\delta - (u_\delta)_{2r}|^3 \, dx \right)$$

$$=: c \sum_{i=1}^{3} I_2^i . \tag{46}$$

Note that this estimate is not sharp in the sense that the constant occurring in the second line in fact tends to zero as $\delta \to 0$. Anyhow, I_2^1 can be absorbed on the left-hand side of (44), whereas I_2^2 is uniformly bounded with respect to δ. I_2^3 is estimated with the help of (43), Hölder's and Sobolev-Poincarè's inequality

$$
\begin{aligned}
I_2^3 &= \delta^{\frac{1}{2}} \int_{B_{2r}(x_0)} |u_\delta - (u_\delta)_{2r}|^3 \, dx \\
&\leq \delta^{\frac{1}{2}} \left(\int_{B_{2r}(x_0)} |u_\delta - (u_\delta)_{2r}|^4 \, dx \right)^{\frac{1}{2}} \\
&\qquad \cdot \left(\int_{B_{2r}(x_0)} |u_\delta - (u_\delta)_{2r}|^2 \, dx \right)^{\frac{1}{2}} \\
&\leq c\delta^{\frac{1}{2}} \int_{B_{2r}(x_0)} \left| \nabla |u_\delta - (u_\delta)_{2r}|^2 \right| \, dx \\
&\leq c\delta^{\frac{1}{2}} \left(\int_{B_{2r}(x_0)} |\nabla u_\delta|^2 \, dx \right)^{\frac{1}{2}} \left(\int_{B_{2r}(x_0)} |u_\delta - (u_\delta)_{2r}|^2 \, dx \right)^{\frac{1}{2}} \\
&\leq c \, .
\end{aligned}
\tag{47}
$$

If we recall that the "ε–terms" are absorbed on the left-hand side of (44) (as described above), then (45)–(47) complete the proof of Lemma 4.33. $\quad\square$

Remark 4.35. *Going through the proof of Lemma 4.33 we see that the assertion is not depending on the exponent μ of ellipticity.*

The above lemma now yields uniform higher local integrability of $|\nabla u_\delta|$ in the scalar case.

Theorem 4.36. *Consider the two-dimensional scalar case $n = 2$, $N = 1$ together with the general assumption of Theorem 4.31. If $B_{2r}(x_0) \Subset \Omega$ then there exists a local constant c, independent of δ, such that*

$$
\int_{B_r(x_0)} \left(1 + |\nabla u_\delta|^2 \right)^{\frac{1}{2}} \ln \left(1 + |\nabla u_\delta|^2 \right) \, dx \leq c \, .
$$

Proof. We let $\omega_\delta = \ln(1 + |\nabla u_\delta|^2)$ and choose $\varphi = (u_\delta - (u_\delta)_{2r}) \omega_\delta \eta^2$, $\eta \in C_0^\infty(B_{2r}(x_0))$, $0 \leq \eta \leq 1$, $\eta \equiv 1$ on $B_r(x_0)$. Again φ is seen to be admissible in the Euler equation (14), Section 4.2.1, and we obtain

$$\int_{B_{2r}(x_0)} \nabla f(\nabla u_\delta)\, \nabla u_\delta\, \omega_\delta\, \eta^2\; dx + \delta \int_{B_{2r}(x_0)} |\nabla u_\delta|^2\, \omega_\delta\, \eta^2\; dx$$

$$= - \int_{B_{2r}(x_0)} \nabla f(\nabla u_\delta)\, \nabla \omega_\delta \left(u_\delta - (u_\delta)_{2r}\right) \eta^2\; dx$$

$$-2 \int_{B_{2r}(x_0)} \nabla f(\nabla u_\delta)\, \nabla\eta\, \eta \left(u_\delta - (u_\delta)_{2r}\right) \omega_\delta\; dx$$

$$-\delta \int_{B_{2r}(x_0)} \nabla u_\delta\, \nabla \omega_\delta \left(u_\delta - (u_\delta)_{2r}\right) \eta^2\; dx$$

$$-2\delta \int_{B_{2r}(x_0)} \nabla u_\delta\, \nabla\eta\, \eta \left(u_\delta - (u_\delta)_{2r}\right) \omega_\delta\; dx$$

$$=: \sum_{i=1}^{4} I_i\; .$$

Similar to the proof of Lemma 4.33, a lower bound for the first integral on the left-hand side is given by Remark 4.2, i), thus

$$\int_{B_{2r}(x_0)} \left(1 + |\nabla u_\delta|^2\right)^{\frac{1}{2}} \omega_\delta\, \eta^2\; dx + \delta \int_{B_{2r}(x_0)} |\nabla u_\delta|^2\, \omega_\delta\, \eta^2\; dx$$

$$\leq c \left(\int_{B_{2r}(x_0)} \omega_\delta\, \eta^2\; dx + \sum_{i=1}^{4} |I_i|\right)\; . \tag{48}$$

Clearly $\int_{B_{2r}(x_0)} \omega_\delta\, \eta^2\; dx$ is uniformly bounded with respect to δ and in order to find an estimate for I_1 we observe

$$|\nabla \omega_\delta|^2 \leq \frac{4}{1 + |\nabla u_\delta|^2} |\nabla^2 u_\delta|^2\; .$$

This, together with Lemma 4.33, implies (again we make use of the fact that $|\nabla f|$ is bounded)

$$|I_1| \leq c \int_{B_{2r}(x_0)} |u_\delta - (u_\delta)_{2r}|\, |\nabla \omega_\delta|\, \eta^2\; dx$$

$$\leq c \left(\int_{B_{2r}(x_0)} \left(1 + |\nabla u_\delta|^2\right)^{\frac{1}{2}} |u_\delta - (u_\delta)_{2r}|^2\, \eta^2\; dx\right)^{\frac{1}{2}}$$

$$\cdot \left(\int_{B_{2r}(x_0)} \left(1 + |\nabla u_\delta|^2\right)^{-\frac{1}{2}} |\nabla \omega_\delta|^2\, \eta^2\; dx\right)^{\frac{1}{2}}$$

$$\leq c \left(\int_{B_{2r}(x_0)} \left(1 + |\nabla u_\delta|^2\right)^{-\frac{3}{2}} |\nabla^2 u_\delta|^2\, \eta^2\; dx\right)^{\frac{1}{2}}\; .$$

Here the right-hand side is bounded through the Caccioppoli-type inequality stated in Lemma 3.19 or Lemma 4.19, i). Note that we exactly reach the limit case $\mu = 3$. Next,

$$|I_2| \leq c \int_{B_{2r}(x_0)} \left(|u_\delta - (u_\delta)_{2r}|^2 + \eta^2 |\nabla \eta|^2 \omega_\delta^2 \right) dx \leq c$$

is immediately verified and

$$|I_3| \leq c\delta \int_{B_{2r}(x_0)} \left(|\nabla u_\delta|^2 |u_\delta - (u_\delta)_{2r}|^2 \eta^2 + |\nabla \omega_\delta|^2 \eta^2 \right) dx$$

$$\leq c \left(1 + \delta \int_{B_{2r}(x_0)} \left(1 + |\nabla u_\delta|^2 \right)^{-1} |\nabla^2 u_\delta|^2 \eta^2 \, dx \right)$$

$$\leq c$$

again follows from Lemma 4.33 and Lemma 4.19, i). Thus, together with

$$|I_4| \leq c\delta \int_{B_{2r}(x_0)} \left(|\nabla u_\delta|^2 |u_\delta - (u_\delta)_{2r}|^2 \eta^2 + |\nabla \eta|^2 \omega_\delta^2 \right) dx \leq c \, ,$$

the theorem is proved recalling (48) and since the constants occurring above are not depending on δ. $\square$

Let us now turn our attention to the vectorial setting $N > 1$.

Theorem 4.37. *Consider the case $N > 1$ and suppose that the assumptions of Theorem 4.31, in particular (2), Section 4.1, hold. Then we still have the claims of Theorem 4.36.*

Remark 4.38. *As outlined in the previous section, this result is combined with Theorem A.11 to obtain a* Proof of Theorem 4.31.

Proof of Theorem 4.37. The theorem is established once the following claims are verified (we keep the notation introduced above).

i) $\varphi = |u_\delta - (u_\delta)_{2r}|^2 (u_\delta - (u_\delta)_{2r}) \eta^2$ is admissible in the Euler equation (14), Section 4.2.1 (this test-function is used to prove Lemma 4.33).

ii) This choice of φ implies (44).

iii) The test-function $\varphi = \left(u_\delta - (u_\delta)_{2r} \right) \omega_\delta \eta^2$ also is admissible (this is necessary for the proof of Theorem 4.36).

If i)–iii) are valid, then the remaining arguments given in the proofs of Lemma 4.33 and Theorem 4.36 can be carried over to the vectorial setting without any changes.

ad i) & iii). We already have noted several times (see, for instance, (28), Section 3.3.3) that u_δ is of class $W^2_{2,loc}(\Omega; \mathbb{R}^N) \cap W^1_{\infty,loc}(\Omega; \mathbb{R}^N)$. This immediately gives i) and iii).

ad *ii*). Here we cannot refer to Remark 4.2 without a further comment since the structure condition is violated if we pass to the energy density $\bar{f} = f(Z) - \nabla f(0) : Z$. However, the representation $f(Z) = g(|Z|^2)$ also implies

$$\nabla f(0) = 0 \, ,$$

and then we may follow the arguments of Remark 4.2. Thus, we obtain $\nabla f(Z) : Z \geq 0$ and (with the notation $f_\delta(Z) = g_\delta(|Z|^2)$)

$$g_\delta'(|Z|^2) \geq 0 \quad \text{for all } Z \in \mathbb{R}^{2N} \, .$$

We next let $\psi = |u_\delta - (u_\delta)_{2r}|^2 \, (u_\delta - (u_\delta)_{2r})$ and have almost everywhere

$$\begin{aligned}
\nabla f_\delta(\nabla u_\delta) : \nabla \psi &= 2 \, g_\delta'(|\nabla u_\delta|^2) \, \nabla u_\delta : \nabla \psi \\
&= 2 \, g_\delta'(|\nabla u_\delta|^2) \left[\partial_\alpha u_\delta^i \, \partial_\alpha u_\delta^i \, |u_\delta - (u_\delta)_{2r}|^2 \right. \\
&\qquad \left. + \left(\partial_\alpha u_\delta^i \left(u_\delta^i - (u_\delta)_{2r}^i \right) \right) 2 \left(\partial_\alpha u_\delta^j \left(u_\delta^j - (u_\delta)_{2r}^j \right) \right) \right] \\
&\geq 2 \, g_\delta'(|\nabla u_\delta|^2) \, \partial_\alpha u_\delta^i \, \partial_\alpha u_\delta^i \, |u_\delta - (u_\delta)_{2r}|^2 \\
&= \nabla f_\delta(\nabla u_\delta) : \nabla u_\delta \, |u_\delta - (u_\delta)_{2r}|^2 \, .
\end{aligned}$$

This implies (44) exactly in the same way as above. □

4.3.2 The case $\mu < 3$

It remains to give a *Proof of Theorem 4.32.* We proceed in three steps:

first we fix a L^1-cluster point $u^* \in \mathcal{M}$ of the regularizing sequence $\{u_\delta\}$ and use Theorem 4.31 to define a suitable local auxiliary variational problem. Here we find uniform local gradient estimates according to Theorem 4.28.

Then, the auxiliary solutions are modified and extended to the whole domain Ω. We obtain a sequence $\{w_m\}$ where it turns out that L^1-cluster points w^* are generalized minimizers of the original problem, hence elements of the set $\mathcal{M}$.

Finally, the duality relation holds almost everywhere both for u^* and for w^* which gives the theorem.

Step 1. From now on we suppose that the assumptions of Theorem 4.32 are valid. We fix a L^1-cluster point u^* of the regularizing sequence $\{u_\delta\}$ and recall that u^* is of class $W_1^1(\Omega; \mathbb{R}^N)$ (compare Section 4.3.1). Moreover, we fix $x_0 \in \Omega$ and write (with a slight abuse of notation) $u^*(r, \theta) = u^*(x_0 + re^{i\theta})$. We assume that $B_{2R_0}(x_0) \Subset \Omega$ and observe that

$$\int_0^{R_0} \int_0^{2\pi} \left| \frac{\partial u^*}{\partial \theta} \right| \, d\theta \, dr \leq \int_0^{R_0} \int_0^{2\pi} |\nabla u^*| \, d\theta \, r \, dr \leq c < \infty \, .$$

Hence, there exists a radius $R_0/2 \leq R \leq R_0$ such that

$$\int_0^{2\pi} \left| \frac{\partial u^*(R, \theta)}{\partial \theta} \right| \, d\theta \leq c < \infty \, . \tag{49}$$

Next we pass to a smooth sequence $\{u_m\}$, $u_m \in C^\infty(\Omega; \mathbb{R}^N)$, with the property

$$u_m \to u^* \quad \text{in } W_1^1(\Omega; \mathbb{R}^N) \text{ as } m \to \infty \, , \tag{50}$$

hence it is possible to estimate

$$\int_0^{R_0} h_m(r) \, dr := \int_0^{R_0} \int_0^{2\pi} \left| \frac{\partial(u_m - u^*)}{\partial \theta} \right| \, d\theta \, dr$$

$$\leq \int_0^{R_0} \int_0^{2\pi} |\nabla(u_m - u^*)| \, d\theta \, r \, dr \xrightarrow{m \to \infty} 0 \, .$$

Thus, $h_m(r) \to 0$ in $L^1((0, R_0))$ as $m \to \infty$ and we may assume in addition to (49) that R is chosen to satisfy

$$\int_0^{2\pi} \left| \frac{\partial u_m(R, \theta)}{\partial \theta} \right| \, d\theta \leq c < \infty \, , \tag{51}$$

where the constant c does not depend on m. As a consequence of (51) it is finally established: there is a radius $R \in (R_0/2, R_0)$ and real number $K > 0$ such that for all $m \in \mathbb{N}$

$$\left| u_{m|\partial B_R(x_0)} \right| \leq K \, . \tag{52}$$

We have found suitable boundary data to consider the variational problem

$$J_\delta[w, B_R(x_0)] := \int_{B_R(x_0)} f(\nabla w) \, dx + \frac{\delta}{2} \int_{B_R(x_0)} |\nabla w|^2 \, dx \tag{$\mathcal{P}_\delta^m$}$$

$$\to \min \quad \text{in } u_m + \overset{\circ}{W}_2^1(B_R(x_0); \mathbb{R}^N) \, .$$

If $\delta = \delta(m)$ is chosen sufficiently small and if we denote by v_m the unique solution of $(\mathcal{P}_\delta^m)$, then

$$J_{\delta(m)}[v_m, B_R(x_0)] \leq J_{\delta(m)}[u_m, B_R(x_0)] \leq c \tag{53}$$

holds with a constant c not depending of m. Moreover, by (52), we find (citing for example the maximum principle given in [DLM] or the convex hull property discussed in the next chapter)

$$\|v_m\|_{L^\infty(B_R(x_0); \mathbb{R}^N)} \leq K \, . \tag{54}$$

At this point we recall that the a priori gradient estimates established in Theorem 4.28 only depend on the data and the constants occurring on the right-hand side of (53) and (54), respectively. As a result, a real number $c > 0$, independent of m, is found such that

$$\|\nabla v_m\|_{L^\infty(B_{R/2}(x_0);\mathbb{R}^{2N})} \leq c. \tag{55}$$

Step 2. Given u^*, u_m and v_m as above, we choose $\eta \in C^\infty(B_R(x_0))$, $\eta \equiv 1$ on $B_R(x_0) - B_{3R/4}(x_0)$, $\eta \equiv 0$ on $B_{R/2}(x_0)$, and let $w_m^1 \colon B_R(x_0) \to \mathbb{R}^N$,

$$w_m^1 := v_m + \eta(u^* - u_m), \quad \text{hence} \quad w_m^1{}_{|\partial B_R(x_0)} = u^*{}_{|\partial B_R(x_0)}.$$

We then claim that w_m^1 provides a $J_{|B_R(x_0)}$-minimizing sequence with respect to the boundary data $u^*_{|B_R(x_0)}$: in fact, (50) implies as $m \to \infty$

$$\left| \int_{B_R(x_0)} \left(f(\nabla u_m) - f(\nabla u^*) \right) dx \right| \leq c \int_{B_R(x_0)} |\nabla u_m - \nabla u^*| \, dx \to 0,$$

and if we decrease δ (if necessary), then we obtain from the minimality of v_m

$$\int_{B_R(x_0)} f(\nabla v_m) \, dx \quad \leq \quad J_{\delta(m)}\big[v_m, B_R(x_0)\big] \leq J_{\delta(m)}\big[u_m, B_R(x_0)\big]$$
$$\overset{m \to \infty}{\to} \int_{B_R(x_0)} f(\nabla u^*) \, dx. \tag{56}$$

Moreover, we have

$$\left| \int_{B_R(x_0)} \left(f(\nabla w_m^1) - f(\nabla v_m) \right) dx \right| \quad \leq \quad c \int_{B_R(x_0)} |\nabla(\eta(u^* - u_m))| \, dx$$
$$\overset{m \to \infty}{\to} 0,$$

which, together with (56) and the minimality of u^* (recall that $u^* \in W_1^1(\Omega; \mathbb{R}^N)$ is a local J-minimizer) implies

$$\int_{B_R(x_0)} f(\nabla w_m^1) \, dx \overset{m \to \infty}{\to} \int_{B_R(x_0)} f(\nabla u^*) \, dx, \tag{57}$$

i.e. our claim is proved.

Now once more Lemma B.5 on local comparison functions comes into play: the trace of each w_m coincides with the trace of u^* on $\partial B_R(x_0)$. Further, we have (57). Thus, it is possible to extend the sequence $\{w_m^1\}$ to a J-minimizing sequence $\{w_m\}$ from $u_0 + \overset{\circ}{W}_1^1(\Omega; \mathbb{R}^N)$. Hence, if w^* denotes a weak cluster point of $\{w_m\}$, then $w^* \in \mathcal{M}$.

Step 3. Finally we recall that u^* is a weak cluster point of the regularizing sequence $\{u_\delta\}$. Hence, the results of Sections 2.3.1 and 2.3.2 apply and we find an open set Ω_0 of full measure such that

$$u^* \in C^{1,\alpha}(\Omega_0; \mathbb{R}^N) \,.$$

Furthermore, if σ denotes the unique solution of the dual variational problem, then

$$\sigma = \nabla f(\nabla u^*) \quad \text{in } \Omega_0 \,.$$

At this point we like to emphasize that the variation of σ leading to Theorem A.9 is of local character. In particular, the assertion remains valid for any generalized minimizer $v^* \in \mathcal{M}$ with Ω replaced by the open set Ω_0, where σ is known to be a continuous function taking values in $\text{Im}(\nabla f)$. Hence we also have

$$\sigma = \nabla f(\nabla v^*) \quad \text{in } \Omega_0 \,.$$

Since $w^* \in \mathcal{M}$ we obtain

$$\nabla w^* = \nabla u^* \quad \text{almost everywhere} \,.$$

On the other hand, recall that

$$w_m|_{B_{R/2}(x_0)} = w_m^1|_{B_{R/2}(x_0)} = v_m|_{B_{R/2}(x_0)} \,,$$

hence the a priori estimate (55) yields

$$\|\nabla u^*\|_{L^\infty(B_{R/2}(x_0); \mathbb{R}^{2N})} \leq c \,.$$

Note u^* really is Lipschitz continuous since $u^* \in W_1^1(\Omega; \mathbb{R}^N)$, in particular $\nabla^s u^* \equiv 0$, is due to the previous considerations. Keeping Remark 4.30 in mind we have proved Theorem 4.32. $\quad \square$

4.4 A counterexample

The studies of linear growth variational problems are completed with an example which shows the sharpness of our results. The idea originates from [GMS1], Example 3.2, where the authors restrict themselves to the one-dimensional situation. We follow the proposal of Giaquinta, Modica and Souček and give a rigorous proof that the arguments extend to higher-dimensional annuli Ω. What is more, the example given in [GMS1] is degenerated which is not the case in the modification outlined below. As a consequence, we precisely can verify the assumptions of Section 4.2.2.2 (note that our arguments rely on a smooth x-dependence) with the exception that we now have $\mu > 3$.

The general setting is the following: let $N = 1$, $n = 2$ and $|x| = \sqrt{x_1^2 + x_2^2} = r$. We fix some positive numbers $0 < \rho_1 < \rho_2$, $\rho := (\rho_1 + \rho_2)/2$ and choose

$$\Omega := \left\{ x \in \mathbb{R}^2 : \rho_1 < r < \rho_2 \right\}.$$

Moreover, $\alpha \colon \Omega \to \mathbb{R}$ is defined by

$$\alpha(r) := 1 + \gamma |r - \rho|^2 ,$$

where the positive parameter γ is chosen later on (see (62) and (73)).

If $k > 2$ is fixed, then the energy density under consideration is given by

$$f(x, P) = f(r, P) = \begin{cases} \left(1 + \alpha(r)|P|^k\right)^{\frac{1}{k}} & \text{if } |P| > \varepsilon , \\[2mm] h(r, P) & \text{if } |P| \le \varepsilon . \end{cases}$$

Here $h(r, P)$ is chosen such that $f(x, P)$ is strictly convex, non-degenerate in P and such that $f(x, P)$ of class $C^2(\overline{\Omega} \times \mathbb{R}^2)$. For an explicit construction we may consider

$$\tilde{f}(P) = \begin{cases} \left(1 + |P|^k\right)^{\frac{1}{k}} & \text{if } |P| > \varepsilon , \\[2mm] \tilde{h}(P) & \text{if } |P| \le \varepsilon , \end{cases}$$

together with the Ansatz

$$\tilde{h}(P) = a + b\left(1 + |P|^l\right)^{\frac{1}{l}} + c|P|^2 ,$$

where a, b and c are suitable constants and $l > k$. The requirement that $\tilde{f}$ is of class $C^2(\mathbb{R}^2)$ in particular implies

$$b = \left(1 + |\varepsilon|^k\right)^{\frac{1}{k} - 2} \left(1 + |\varepsilon|^l\right)^{2 - \frac{1}{l}} |\varepsilon|^{k-l} \frac{k - 2 - |\varepsilon|^k}{l - 2 - |\varepsilon|^l} > 0 ,$$

$$c = \frac{1}{2} |\varepsilon|^{k-2} \left(1 + |\varepsilon|^k\right)^{\frac{1}{k} - 2} \left[\left(1 + |\varepsilon|^k\right) - \left(1 + |\varepsilon|^l\right) \frac{k - 2 - |\varepsilon|^k}{k - 2 - |\varepsilon|^l} \right] > 0 .$$

We then let $h(r, P) := \tilde{h}([\alpha(r)]^{1/k} P)$.

Finally, the choice of the second parameter $0 < \varepsilon < 1$ will be made in the inequality (70) below.

Theorem 4.39. *With the above notation, the variational problem*

$$J[w] := \int_{\Omega} f(x, \nabla w) \, dx \to \min \quad in \ u_0 + \overset{\circ}{W}{}_1^1(\Omega)$$

does not admit a generalized minimizer $v \in \mathcal{M}$ of class $W_1^1(\Omega)$ if u_0 is supposed to satisfy $u_0(\rho_1) = -a$ and $u_0(\rho_2) = a$ for a constant $a > 0$ sufficiently large (see (74)). Here and in the following – again with a slight abuse of notation – we write $u(x_1, x_2) = u(r)$ whenever u is merely depending on $|x|$.

Remark 4.40.

i) *Note that the ellipticity exponent of f is given by $\mu = k + 1 > 3$, hence we really obtain an example on the sharpness of our results.*

ii) *Moreover, it should be emphasized that the boundary values u_0 may be chosen as a function of class $C^\infty(\overline{\Omega})$.*

Proof of Theorem 4.39. Assume by contradiction that $v \in W_1^1(\Omega)$ is a generalized minimizer. Then the proof of our assertion splits into three steps.

Step 1. First of all we note that, by the symmetry of the problem and with the obvious meaning of notation (after introducing polar coordinates), we have

$$v(r, \varphi) = v(r) . \tag{58}$$

In fact, consider the regularization $\{u_\delta\}$ of Section 4.2 which clearly satisfies $u_\delta(r, \varphi) = u_\delta(r)$ since for any real number φ_0 the function $u_\delta(r, \varphi + \varphi_0)$ is J_δ-minimizing with respect to the boundary data u_0 as well. Hence, the uniqueness of minimizers proves the claim for u_δ. Now recall the reasoning of Section 4.3.2, Step 3, to see that almost everywhere

$$\nabla u^* = \nabla^a u^* = \nabla v = \nabla^a v ,$$

where $u^* = u^*(r)$ denotes a L^1-cluster point of the sequence $\{u_\delta\}$. Thus we have (58) recalling $v \in W_1^1(\Omega)$.

Step 2. We next claim that v takes the boundary data u_0 in the sense that the trace of v on $\partial\Omega$ is just u_0, i.e.

$$v(\rho_1) = -a \quad \text{and} \quad v(\rho_2) = a . \tag{59}$$

In order to prove (59) we consider the comparison function

$$w(r) = \begin{cases} v(r) - v(\rho_1) - a & \rho_1 < r < \rho , \\[2mm] v(r) - v(\rho_2) + a & \rho \leq r < \rho_2 , \end{cases}$$

and assume by contradiction that (59) is violated. If we observe that

$$f_\infty(r, P) = \alpha^{\frac{1}{k}}(r)|P| , \quad \rho_1 < r < \rho_2 , \ P \in \mathbb{R}^2 ,$$

then we obtain with the notation of Appendix A

$$K[w] = \int_\Omega f(r, \nabla^a w) \, dx + \int_{\partial B_\rho(0)} d|\nabla^s w| . \tag{60}$$

Here we used the fact that $\nabla^s w$ is supported on $\partial B_\rho(0)$ and that w takes its boundary data u_0 on $\partial\Omega$. From [AFP], Theorem 3.77, p. 171, one gets

$$\nabla^s w \llcorner \partial B_\rho(0) = \big(v(\rho_1) - v(\rho_2) + 2a\big) \frac{x}{|x|} \, d\mathcal{H}^1 \, .$$

Thus, (60) may be written in the form

$$
\begin{aligned}
K[w] &= \int_\Omega f(r, \nabla^a w) \, dx + 2\pi\rho \big|v(\rho_1) - v(\rho_2) + 2a\big| \\
&\leq \int_\Omega f(r, \nabla v) \, dx + 2\pi\rho \Big(\big|v(\rho_1) + a\big| + \big|v(\rho_2) - a\big|\Big) \, .
\end{aligned}
\tag{61}
$$

Now choose γ sufficiently large such that for $i = 1, 2$

$$\frac{\rho}{\rho_i \, \alpha^{\frac{1}{k}}(\rho_i)} < 1 \, . \tag{62}$$

Then we obtain

$$K[w] < \int_\Omega f(r, \nabla v) \, dx + \sum_{i=1}^{2} \alpha^{\frac{1}{k}}(\rho_i) \int_{\partial B_{\rho_i}(0)} |u_0 - v(\rho_i)| \, d\mathcal{H}^1 = K[v] \, ,$$

hence the desired contradiction (recall Theorem A.3, *iii*), which remains valid with an additional smooth x-dependence.

Step 3. Now we make use of the Euler equation for the generalized minimizer v which takes the standard form since v is assumed to be of class $W_1^1(\Omega)$, i.e. we have

$$\int_\Omega \nabla_P f(r, \nabla v) \cdot \nabla \psi \, dx = 0 \quad \text{for all } \psi \in C_0^1(\Omega) \, . \tag{63}$$

In particular, this is true for test-functions $\psi = \psi(r) \in C_0^1((\rho_1, \rho_2))$. In the following the derivative with respect to r is denoted by "$\cdot$". Then, again using polar coordinates, $\nabla v = (\cos\varphi \dot{v}, \sin\varphi \dot{v})$, and with the notation $\nabla_P f(r, \nabla v) = g(r, |\dot{v}|)\nabla v$ we obtain from (63)

$$\int_{\rho_1}^{\rho_2} \int_0^{2\pi} g(r, |\dot{v}|) \dot{v} \dot{\psi} r \, dr \, d\varphi = 0 \quad \text{for all } \psi = \psi(r) \in C_0^1((\rho_1, \rho_2)) \, .$$

As a consequence, there is a real number $\lambda \in \mathbb{R}$ such that for all $r \in (\rho_1, \rho_2)$

$$g(r, |\dot{v}|) \dot{v} r = \lambda \, . \tag{64}$$

With the representation

$$
g(r, |\dot{v}|) =
\begin{cases}
\big(1 + \alpha(r)|\dot{v}|^k\big)^{\frac{1}{k}-1} \alpha(r)|\dot{v}|^{k-2} & \text{if } |\dot{v}| > \varepsilon \, , \\[2mm]
b\big(1 + \alpha(r)|\dot{v}|^l\big)^{\frac{1}{l}-1} \alpha(r)|\dot{v}|^{l-2} + 2c[\alpha(r)]^{\frac{2}{k}} & \text{if } |\dot{v}| < \varepsilon
\end{cases}
$$

we have to distinguish two cases.

Case 1. If $|\dot{v}| < \varepsilon$, then using the formulas for b and c we immediately see that $g(r, |\dot{v}|) \leq c|\varepsilon|^{k-2}$, in particular

$$\varepsilon > |\dot{v}| = \frac{|\lambda|}{g(r, |\dot{v}|) r} \geq c|\lambda| \tag{65}$$

for some positive constant c.

Case 2. If $|\dot{v}| > \varepsilon$, then (64) implies by elementary calculations (note that $|\lambda| > 0$ in the case at hand)

$$|\dot{v}|^{-k} \alpha^{-\frac{k}{k-1}} + \alpha^{-\frac{1}{k-1}} = (|\lambda|/r)^{-\frac{k}{k-1}} . \tag{66}$$

Observe that, as a consequence of (66),

$$(|\lambda|/r)^k \leq \alpha(r) . \tag{67}$$

Now, again with some simple computations, (66) gives

$$|\dot{v}| = \frac{(|\lambda|/r)^{\frac{1}{k-1}}}{\alpha^{\frac{1}{k}} \left(\alpha^{\frac{1}{k-1}} - (|\lambda|/r)^{\frac{k}{k-1}} \right)^{\frac{1}{k}}} . \tag{68}$$

Summarizing both cases we have the formulas (65) and (68) for $|\dot{v}|$.

We then choose $\lambda_0 > 0$ sufficiently small such that (68) (which is independent of the parameter $\varepsilon > 0$) implies $|\dot{v}| \leq 1$ and assume that $|\lambda| < \lambda_0$. Then, by (65) and (68) we see that $|\dot{v}| \leq 1$ for all $r \in (\rho_1, \rho_2)$. On the other hand, v takes its boundary data and $v(r)$ is of class $W_1^1((\rho_1, \rho_2))$, hence $v(r)$ is an absolutely continuous function and we may write

$$a = \frac{1}{2} |v(\rho_1) - v(\rho_2)| \leq \frac{1}{2} \int_{\rho_1}^{\rho_2} |\dot{v}(r)| \, dr \leq \frac{\rho_2 - \rho_1}{2} . \tag{69}$$

This gives a contradiction if a is sufficiently large and we may assume $|\lambda| \geq \lambda_0$ which was chosen independent of ε. Hence, if Case 1 holds true, then (65) yields

$$\varepsilon \geq c|\lambda_0| , \tag{70}$$

and we choose ε sufficiently small such that this is not possible. Once it is established that Case 2 holds for all $r \in (\rho_1, \rho_2)$, we obtain from (67)

$$|\lambda| \leq \inf_{r \in (\rho_1, \rho_2)} \alpha^{\frac{1}{k}}(r) r . \tag{71}$$

Moreover, (68) gives the right representation and using (67), (71), $\alpha \geq 1$ and $k > 2$ we estimate

$$|\dot{v}| \leq \frac{\alpha^{\frac{1}{k(k-1)}}}{\alpha^{\frac{1}{k}}\left(\alpha^{\frac{1}{k-1}} - (|\lambda|/r)^{\frac{k}{k-1}}\right)^{\frac{1}{k}}} \leq \frac{\alpha^{\frac{2-k}{k(k-1)}}}{\left(\alpha^{\frac{1}{k-1}} - (|\lambda|/r)^{\frac{k}{k-1}}\right)^{\frac{1}{k}}} \tag{72}$$

$$\leq \left[\alpha^{\frac{1}{k-1}} - \left(r^{-1}\inf_{r\in(\rho_1,\rho_2)}[\alpha^{1/k}(r)r]\right)^{\frac{k}{k-1}}\right]^{-\frac{1}{k}} =: [h(r)]^{-\frac{1}{k}}.$$

Here we first note that $h(r)$ is independent of λ, in particular $h(r)$ does not depend on the boundary values u_0 given in terms of a. Moreover, $h(r) \geq 0$ is evident by definition. Finally, the zeros of $h(r)$ are of finite number and simultaneously (by (62) interior) minima of $h(r)$. With Taylor's formula we obtain the expansion

$$h(r) \approx c(r - r_0)^2 \quad \text{near the zeros } r_0 \text{ of } h(r),$$

since $h''(r_0)$ is not vanishing. In fact, assume by contradiction that

$$h(r_0) = 0, \quad h'(r_0) = 0, \quad h''(r_0) = 0.$$

This leads (by elementary calculations) to

$$r_0^{2+k} \leq \frac{k(1+k)\rho^k}{2\gamma}, \tag{73}$$

hence if $\gamma = \gamma(k, \rho_1, \rho_2)$ is chosen sufficiently large, then no such zeros are possible for the radii under consideration. With the above expansion we may choose $a < \infty$ such that

$$\int_{\rho_1}^{\rho_2} [h(r)]^{-\frac{1}{k}}\, dr < a. \tag{74}$$

This proves the theorem since (72) and (74) contradict (69). □

Remark 4.41. *Let us again concentrate on the regularizing sequence $\{u_\delta\}$ with L^1-cluster point u^* as studied in the previous sections. Then it is not difficult to locate the singular set of u^* in the situation at hand. To this purpose denote by $\rho_{0,i}$, $i = 1, \ldots, M$, the minima (of finite number, lying in the interior of (ρ_1, ρ_2) by (62)) of the function $\alpha^{\frac{1}{k}}(r)r$ on (ρ_1, ρ_2). We then have*

$$\operatorname{spt} \nabla^s u^* \subset \bigcup_{i=1}^{M} \partial B_{\rho_{0,i}}(0). \tag{75}$$

In fact, the sequence of radially symmetric functions $\{u_\delta\} = \{u_\delta(r)\}$ yields a minimizing sequence of the one-dimensional energy ($\Omega = I = (\rho_1, \rho_2)$)

$$J_I[w] := \int_I f(r, |\dot{w}(r)|)\, r\, dr$$

with respect to $W_1^1(I)$-comparison functions $w(\rho_1) = -a$, $w(\rho_2) = a$. In this sense u^* provides a generalized J_I-minimizer. Now we again extend the ideas of [GMS1] and let $\dot{u}_a^*$, $\dot{u}_s^*$ denote the Lebesgue-decomposition of $\dot{u}^*$ in the absolutely continuous and singular part, respectively. Moreover, Corollary 3.33, [AFP], p. 140, on the decomposition of functions of bounded variation defined on intervals allows us to choose $\tilde{v}(r) \in W_1^1(I)$ such that for almost all $r \in I$

$$\dot{\tilde{v}}(r) = \dot{u}_a^*(r) \quad \text{and} \quad \tilde{v}(\rho_1) = u^*(\rho_1) .$$

Next let $v(r)$ differ from $\tilde{v}(r)$ just by additional jumps at the points $\rho_{0,i}$ such that

$$\dot{v}_s(r) = \frac{1}{M} \sum_{i=1}^{M} \delta_{\rho_{0,i}} \int_I \dot{u}_s^* ,$$

where $\delta_{\rho_{0,i}}$ denotes the Dirac-measure centered at $\rho_{0,i}$, $i = 1, \ldots, M$. Note that

$$\int_I \dot{u}_a^*(t) \, \mathrm{d}t + \int_I \dot{u}_s^* = u^*(\rho_2) - u^*(\rho_1) ,$$

$$\int_I \dot{v}_a(t) \, \mathrm{d}t + \left(\int_I \dot{u}_s^* \right) \frac{1}{M} \sum_{i=1}^{M} \int_I \delta_{\rho_{0,i}} = v(\rho_2) - v(\rho_1)$$

also implies $v(\rho_2) = u^*(\rho_2)$. Thus we obtain

$$\int_I f(r, |\dot{u}_a^*(r)|) r \, \mathrm{d}r + \int_I \alpha^{\frac{1}{k}}(r) r \, d|\dot{u}_s^*|(r)$$

$$\leq \int_I f(r, |\dot{v}_a(r)|) r \, \mathrm{d}r + \int_I \alpha^{\frac{1}{k}}(r) r \, d|\dot{v}_s|(r)$$

$$\leq \int_I f(r, |\dot{u}_a^*(r)|) r \, \mathrm{d}r + \min_{r \in (\rho_1, \rho_2)} \alpha^{\frac{1}{k}}(r) r \int_I |\dot{u}_s^*| \leq K[u^*]$$

and our claim (75) is proved. $\square$

Remark 4.42. Although u^* as discussed above is not of class $W_1^1(\Omega)$ and although we do not know whether u^* is of class $C^{2,\alpha}$ on the complement of the singular set we might conjecture that there exist analogous examples in the case $\mu = 3$ providing W_1^1-minimizers of

$$\int_\Omega f(\nabla w) \, \mathrm{d}x + \int_{\partial\Omega} f_\infty((u_0 - w)\nu) \, d\mathcal{H}^{n-1} \to \min ,$$

which are smooth on the complement of a finite number of interior spheres. However, if solutions of this kind exist, then they are due to the non-convexity of Ω. In fact, consider a smooth convex domain Ω, assume that $n \geq 2$, $N = 1$, and suppose that there is a $W_1^1(\Omega)$-solution which is of class $C^{2,\alpha}$ near the boundary $\partial\Omega$. Then, on account of the uniqueness of solutions (up to a constant), we apply Hilbert-Haar arguments (compare [MM]) to see that the singular set is empty. In this sense, as the typical behavior, singularities must concentrate near the boundary.

Remark 4.43. *In order to show rigorously that our regularity theory breaks down if $\mu > 3$ we have to ensure that the energy density f studied in Theorem 4.39 satisfies Assumption 4.22 (of course now with ellipticity exponent $\mu = k + 1$). Here it is clearly sufficient to consider $(P \in \mathbb{R}^n)$*

$$f(P) = \left(1 + |P|^k\right)^{\frac{1}{k}}, \quad k > 2,$$

and to study Assumption 4.22 with respect to

$$\tilde{f}(x, P) = f\big(\alpha(x) P\big), \quad \alpha(x) = \left(1 + |x|^2\right)^{\frac{1}{k}},$$

whenever $|P| > 1$ and $x \in B_1(0) \subset \mathbb{R}^n$. To this purpose we first observe that direct calculations yield in the case $|P| > 1$

$$D^2 f(P) \cdot P = \left(1 + |P|^k\right)^{\frac{1}{k}-2} (k-1)|P|^{k-2} P \qquad (76)$$

and, as a direct consequence,

$$\left|D^3 f(P)(P, U, V)\right| \leq c\left|D^2 f(P)(U, V)\right| + c\left(1 + |P|^2\right)^{-\frac{1+k}{2}} |U||V| \qquad (77)$$

for all $U, V \in \mathbb{R}^n$. For the discussion of $\tilde{f}$ we just have to verify iv), v) and vi) of Assumption 4.22, where iv) immediately follows from (76). Now, note that for $1 \leq \gamma \leq n$ and $|P| > 1$

$$\partial_\gamma \partial_\gamma \nabla_P \tilde{f}(x, P) = \partial_\gamma \partial_\gamma \alpha(x) \nabla f(\alpha(x) P)$$

$$+ 2 \partial_\gamma \alpha(x) D^2 f(\alpha(x) P) \cdot P \partial_\gamma \alpha(x)$$

$$+ \alpha(x) D^3 f(\alpha(x) P)(P, P) \left[\partial_\gamma \alpha(x)\right]^2$$

$$+ \alpha(x) D^2 f(\alpha(x) P) \cdot P \partial_\gamma \partial_\gamma \alpha(x) =: \sum_{i=1}^{4} I_i.$$

Clearly I_1 is uniformly bounded and the same follows for I_2 and I_4 from (76). I_3 is estimated with the help of (77), i.e.

$$\left|D^3 f(\alpha(x) P)(P, P)\right| \leq \frac{1}{\alpha(x)} \left|D^3 f(\alpha(x) P)(\alpha(x) P, P)\right|$$

$$\leq c\left|D^2 f(\alpha(x) P)\right| |P| + c\left(1 + |P|^2\right)^{-\frac{k}{2}} \leq c,$$

hence we have v). Finally vi) is established by

$$\partial_\gamma D_P^2 \tilde{f}(x, P)(U, V) = 2\alpha(x) \partial_\gamma \alpha(x) D^2 f(\alpha(x) P)(U, V)$$

$$+ \left[\alpha(x)\right]^2 D^3 f(\alpha(x) P)(P, U, V) \partial_\gamma \alpha(x)$$

if we once more recall (77). $\square$

Bounded solutions for convex variational problems with a wide range of anisotropy

Once more the context of linear growth problems is left. However, we keep in mind that in Section 4.2 – in contrast to Chapter 3 – an application of Sobolev's inequality was replaced by the study of an additional comparison function. This leads to much better results if we impose some uniform boundedness condition on the regularization (see Assumption 4.11). Recall that (up to a certain extent) the limit case $\mu = 3$ is reached and that we cannot expect regular solutions if the ellipticity is given in terms of $\mu > 3$.

With these facts, the conjecture for variational problems with non-standard, superlinear growth is evident: if we impose some boundedness condition as above, then, as a formal correspondence, the relation $1 < q < 4 - \mu$ (for anisotropic power growth integrands $1 < q < 2 + p$) is expected to be the best possible one inducing (partially) regular solutions. Note that the relevance of the restriction $q < 2 + p$ was already discovered in [ELM2]: given a uniform L^∞-regularization $\{u_\delta\}$, uniform local higher (up to a certain extent) integrability of $|\nabla u|$ is established in the vector-valued setting (choosing $2 \leq p$).

Nevertheless, the full strength of the above stated correspondence could not be established in the paper [BF5] on anisotropic variational integrals with convex hull property: instead of $1 < q < 2 + p$ (plus the condition $q < pn/(n-2)$ of Section 3.4 improving higher integrability to partial regularity) the exponents have to be related by $1 < q < p + 2/3$. This is caused by an essential difference to the linear growth situation: in Section 4.2 we benefit from the growth rate $1 = q$ of the main quantity $\nabla f(Z) : Z$ under consideration. Given an anisotropic power growth integrand, we just have the lower bound $p < q$ of this quantity.

As a consequence, the techniques again have to be changed such that we do not have to rely on the quantity $\nabla f(Z) : Z$. This leads to the study of Choe's article [Ch], where bounded solutions with respect to integrands which are of non-standard growth and which are of the form $f(Z) = g(|Z|^2)$ (an assumption both for $N > 1$ and the scalar situation) are handled up to $1 < q < p + 1$. As a third approach, his results depend on an integration by parts combined with a Caccioppoli-type inequality which is slightly different from the one given in Lemma 3.19 or Lemma 4.19, i), respectively.

In this Chapter we are interested in the question whether Choe's Ansatz can be improved. The main idea is to combine some of his arguments with the Caccioppoli-type inequalities as used throughout our whole studies. This gives surprisingly strong results in the sense that the above stated conjecture in fact is valid.

This time we start with the vector-valued case in Section 5.1. The consideration of the scalar case in Section 5.2 then slightly leaves the main line: in the previous sections on scalar problems the structured vectorial case is included without essential modifications. This well known reasoning is omitted in the following for the sake of simplicity. Here we want to emphasize two other aspects (which similarly can be included in the previous studies): on one hand it is shown that the study of obstacle problems follows more or less the same ideas as given in the unconstrained case provided that we incorporate a suitable cut-off function. As a consequence, it is sufficient to prove a priori estimates on level-sets. This, on the other hand, allows us to admit some kind of degeneracy. Moreover, we can formally include and substantially extend the energy densities covered in [Ch].

5.1 Vector-valued problems

Following the existence theory sketched in Section 3.1 (see Theorem 3.3) we fix the unique solution u of the problem

$$J[w] := \int_\Omega \nabla f(\nabla w) \, dx \to \min \quad \text{in } \mathbb{K}_F , \qquad (\mathcal{P})$$

where F is a N-function having the Δ_2-property and where again

$$c_1 F(|Z|) - c_2 \le f(Z) \quad \text{for all } Z \in \mathbb{R}^{nN} \qquad (1)$$

is supposed to hold with some positive constants c_1, c_2. The boundary data u_0 are supposed to satisfy

$$u_0 \in \mathbb{J}_F \cap L^\infty(\Omega; \mathbb{R}^N) . \qquad (2)$$

Our precise assumptions on the energy density f read as

Assumption 5.1. *The energy density $f: \mathbb{R}^{nN} \to \mathbb{R}$ is a function of class $C^2(\mathbb{R}^{nN})$ and its second derivative is estimated for all $Z, Y \in \mathbb{R}^{nN}$ by*

$$\lambda \left(1 + |Z|^2\right)^{-\frac{\mu}{2}} |Y|^2 \le D^2 f(Z)(Y, Y) \le \Lambda \left(1 + |Z|^2\right)^{\frac{q-2}{2}} |Y|^2 . \qquad (3)$$

Here λ, Λ denote some positive constants and the exponents $\mu \in \mathbb{R}$, $q > 1$, are related by

$$q < 4 - \mu \, . \tag{4}$$

Moreover, the representation formula

$$f(Z) = g\big(|Z_1|, \dots, |Z_n|\big) \, , \quad Z = (Z_1, \dots, Z_n) \in \mathbb{R}^{nN} \, , \tag{5}$$

is supposed to be valid for some function g which is increasing with respect to each argument. Finally, we assume the lower bound (1).

Remark 5.2. *Let us shortly discuss the condition (5).*

i) *It is proved in [BF5] that (5) implies the convex hull property: if $\mathrm{Im}(u_0) \subset K$ for a compact convex set $K \subset \mathbb{R}^N$, then the solution u (together with its regularization) also satisfies $\mathrm{Im}(u) \subset K$. In fact, let $\pi \colon \mathbb{R}^N \to K$ denote the projection onto K, hence $\mathrm{Lip}(\pi) = 1$. Then Lemma B.1 of [BF5] proves $|\partial_\gamma(\pi \circ u)| \le |\partial_\gamma u|$, $\gamma = 1, \dots, n$. As a result, the comparison function $v := \pi \circ u$ is minimizing,*

$$g\big(|\partial_1 v|, \dots, |\partial_n v|\big) \le g\big(|\partial_1 u|, \dots, |\partial_n u|\big) \, ,$$

and the uniqueness of solutions proves the claim.

ii) *Condition (5) can be replaced by any other assumption which gives an appropriate maximum principle (once more compare [DLM]). For instance we could follow the lines of Assumption 4.11 or Remark 4.12. In the situation at hand we prefer the formulation (5) as a natural approach to anisotropic variational problems.*

iii) *With the above remark it is again evident that the expression "bounded solutions" occurring in the heading of this chapter is meant in the sense of "uniformly bounded regularizations".*

Let us give some further comments on the choice of f which will be the same in the scalar situation.

Remark 5.3.

i) *In contrast to the notion of (s, μ, q)-growth there is no need to specify the growth rate s of the variational integrand in the following considerations. We just have the bounds induced by (3) (see ii) below).*

ii) *The discussion of Remark 4.2 shows: if $\mu < 1$, then ellipticity is good enough to improve (1) to the power growth estimate (with suitable constants c_1, $c_2 > 0$)*

$$c_1 |Z|^{2-\mu} - c_2 \le f(Z) \quad \text{for all } Z \in \mathbb{R}^{nN} \, .$$

Moreover, f is at most of growth rate q (recall Remark 3.5, iii)).

iii) *Note that (4) – in complete accordance with Section 4.2 – implies*

$$\mu < 3 \, .$$

iv) It should be kept in mind that anisotropic power growth examples (like discussed in Examples 5.7 below) satisfy (4) whenever

$$q < 2 + p.$$

Now the main result of this section is formulated:

Theorem 5.4. *Let u denote the unique solution of the problem (P) where f satisfies Assumption 5.1 and where the boundary values u_0 are given according to (2). Then for any $q < s < 4 - \mu$ and for any $\widehat{\Omega} \Subset \Omega$ there is a positive number c such that*

$$\int_{\widehat{\Omega}} |\nabla u|^s \, dx \le c < \infty.$$

The following comments will also remain unchanged in the scalar case.

Remark 5.5.

i) A corresponding result is true for local minimizers of the energy $J[w]$.

ii) In terms of anisotropic integrands with (p, q)-growth, higher integrability was established in Lemma 3.17 whenever

$$q < p \frac{n+2}{n}, \tag{a}$$

and there was no need to use L^∞-bounds for the solution. Hence, at first glance one may wonder about the case

$$2 + p < p \frac{n+2}{n}, \tag{b}$$

since then the hypothesis $q < 2 + p$ implies (a). Thus, higher integrability holds without an additional boundedness condition. But (b) is equivalent to $p > n$, hence, by Sobolev's embedding theorem, boundedness becomes no restriction at all.

iii) With ii) it is clear that the results extend to the case

$$q < \max\left\{4 - \mu, (2 - \mu)\frac{n+2}{n}\right\}.$$

iv) We do not want to state a conjecture on the sharpness of our results. Nevertheless, recalling Section 4.4 together with the above mentioned correspondences and having the discussion of Section 3.5 in mind, we at least note that our assumptions are reasonable and consistent.

Given Theorem 5.4 we obtain the following corollary on partial regularity. Here we follow the blow-up arguments of Section 3.4.3 which remain unchanged once a Caccioppoli-type inequality and higher local integrability of the gradient are verified. Some more details are outlined in [BF5], let us just note (see Remark 5.8 below) that the way of regularizing the problem is irrelevant since these ingredients are formulated in terms of the solution u.

The restriction

$$q < (2 - \mu)\frac{n}{n-2} \quad \text{if } n \geq 3 \qquad (*)$$

is due to the needed properties of the auxiliary functions ψ_m which were introduced in Section 3.4.3.3. Since our boundedness condition does not improve the Caccioppoli-type inequality, which in return is the basis of the discussion of ψ_m, we cannot expect to get rid of assumption $(*)$.

Corollary 5.6. *The hypotheses of Theorem 5.4 together with the condition*

$$q < \min\left\{4 - \mu, (2 - \mu)\frac{n}{n-2}\right\} \quad \text{if } n \geq 3$$

yield an open set $\Omega_0 \subset \Omega$ of full measure, $|\Omega - \Omega_0| = 0$, such that $u \in C^{1,\alpha}(\Omega_0; \mathbb{R}^N)$ for any $0 < \alpha < 1$.

Before we are going to prove Theorem 5.4 let some examples illustrate our result.

Example 5.7.

i) As a first example we consider the anisotropic energy density

$$f(Z) = \left(1 + |Z|^2\right)^{\frac{p}{2}} + \left(1 + |Z_n|^2\right)^{\frac{q}{2}}, \quad Z = (Z_1, \ldots, Z_n) \in \mathbb{R}^{nN},$$

with exponents $2 \leq p < q$. This structure is imposed in [AF4] to obtain partial regularity under a rather weak condition relating p and q (see [BF5] for a detailed comparison with the results of [PS], [BF5] and Section 3.4).

ii) If we do not have the above decomposition, for instance $(2 \leq p < q)$

$$f(Z) = \left[1 + |Z|^2 + \left(1 + |Z_n|^2\right)^{\frac{q}{p}}\right]^{\frac{p}{2}}, Z = (Z_1, \ldots, Z_n) \in \mathbb{R}^{nN},$$

or if the energy density is completely anisotropic in the sense that

$$f(Z) = \sum_{i=1}^{n} \phi_i(Z_i), \quad \phi_i = \left(1 + |Z_i|^2\right)^{\frac{q_i}{2}}, \quad Z_i \in \mathbb{R}^N,$$

with exponents $q_i \geq 2$, then the results of [AF4] do not apply any more. However, with the notation introduced in (3) (letting $p = 2 - \mu$), partial regularity follows from [BF5] if $q < pn/(n-2)$ and $q < \hat{q} := \max\{p + 2/3; p(n+2)/n\}$. In the following we will improve $p + 2/3$ to $p + 2$.

iii) Let us finally discuss an example which is the most interesting one from our point of view. Let $Z = (Z_1, Z_2) \in \mathbb{R}^{kN} \times \mathbb{R}^{(n-k)N}$, $1 \leq k < n$. Moreover, suppose we are given exponents $1 < p < q < 2$ and

$$f(Z) = \left(1 + |Z|^2\right)^{\frac{p}{2}} + \left(1 + |Z_2|^2\right)^{\frac{q}{2}}.$$

In this subquadratic case (by elementary calculations) the estimate

$$\lambda\left(1+|Z|^2\right)^{\frac{p-2}{2}}|Y|^2 \le D^2 f(Z)(Y,Y) \le \Lambda |Y|^2$$

is the best possible one. As a consequence, no regularity results are available up to now if p is close to 1 – even if $(q-p)$ becomes very small. Hence, with the trivial inequality $2 < p+2$, our theorem really covers a new class of variational integrals.

We now come to the *Proof of Theorem 5.4* and we assume that all the hypotheses stated there are valid. We denote by $(u)_\varepsilon$ the ε-mollification of the minimizer u under consideration through a family of smooth mollifiers, we fix a ball $B := B_R(x_0) \Subset \Omega$ and assume that $B \subset \{x \in \Omega : \text{dist}(x, \partial\Omega) > \varepsilon\}$ for any small $\varepsilon > 0$ as above. Moreover, we choose some exponent $t > \max\{2, q\}$ and let for any $\delta \in (0,1)$

$$f_\delta(Z) := f(Z) + \delta\left(1+|Z|^2\right)^{\frac{t}{2}}, \quad Z \in \mathbb{R}^{nN}.$$

Then u_δ $(= u_\delta^\varepsilon)$ is defined as the unique solution of the Dirichlet problem

$$J_\delta[w, B] := \int_B f_\delta(\nabla w)\,\mathrm{d}x \to \min, \quad w \in (u)_{\varepsilon|B} + \overset{\circ}{W}{}^1_t(B; \mathbb{R}^N). \qquad (\mathcal{P}_\delta)$$

Remark 5.8. *In Remark 3.13 we summarized the reasons why the regularization of Chapter 3 was done with respect to $t = q$. The technique outlined below relies on the condition $t > \max\{2, q\}$: the discussion of asymptotically regular integrands (compare [CE] or [GiaM], Theorem 5.1) includes the vectorial case and yields*

$$u_\delta \in W^1_{\infty,loc}(B; \mathbb{R}^N) \cap W^2_{2,loc}(B; \mathbb{R}^N). \qquad (6)$$

If $\delta = \delta(\varepsilon)$ is chosen sufficiently small (recall the conclusion of Chapter 3.3), then the counterpart of Lemma 3.28 reads as

Lemma 5.9. *With the above notation we have*

i) $\|u_{\delta(\varepsilon)}\|_{W^1_F(B;\mathbb{R}^N)} \le const < \infty;$

ii) $u_{\delta(\varepsilon)} \to u$ *in* $W^1_1(B; \mathbb{R}^N)$ *and almost everywhere as* $\varepsilon \to 0;$

iii) $\sup\limits_B |u_\delta(\varepsilon)| \le \sup\limits_{B_{R+\varepsilon}(x_0)} |u| < \infty;$

iv) $\delta(\varepsilon) \int_B \left(1+|\nabla u_{\delta(\varepsilon)}|^2\right)^{\frac{t}{2}}\,\mathrm{d}x \to 0$ *as* $\varepsilon \to 0;$

v) $\int_B f(\nabla u_{\delta(\varepsilon)})\,\mathrm{d}x \to \int_B f(\nabla u)\,\mathrm{d}x$ *as* $\varepsilon \to 0;$

vi) $\int_B f_{\delta(\varepsilon)}(\nabla u_{\delta(\varepsilon)})\,\mathrm{d}x \to \int_B f(\nabla u)\,\mathrm{d}x.$

Keeping Lemma 5.9 in mind, we again use the notation "$\delta \to 0$" instead of "$\varepsilon \to 0$ and $\delta(\varepsilon)$ sufficiently small" (compare Lemma 3.28). Finally let us recall the Caccioppoli-type inequality (see Lemma 3.19)

Lemma 5.10. *There is a real number $c > 0$, independent of δ, such that for any $\eta \in C_0^\infty(B)$, $0 \le \eta \le 1$,*

$$\int_B D^2 f_\delta(\nabla u_\delta)(\partial_\gamma \nabla u_\delta, \partial_\gamma \nabla u_\delta)\eta^2 \, dx$$

$$\le c \int_B |D^2 f_\delta(\nabla u_\delta)| \, |\nabla u_\delta|^2 \, |\nabla \eta|^2 \, dx .$$

Remark 5.11. *Note that this time the proof is standard: on account of (6) we may differentiate the Euler equation and take $\varphi = \eta^2 \partial_\gamma u_\delta$ as an admissible test-function.*

We obviously (see Lemma 5.9) have established Theorem 5.4 once uniform local higher integrability of the regularization is proved in the sense of

Theorem 5.12. *With the above stated hypotheses, for any $q < s < 4 - \mu$ and for any ball $B_r(x_0)$, $r < R$, there is a constant c just depending on the data, on $\sup_B |(u)_\varepsilon|$ and on r and s, such that*

$$\int_{B_r(x_0)} |\nabla u_\delta|^s \, dx \le c < \infty .$$

Proof. If s is fixed as above then it is possible to define

$$q + \mu - 4 < \alpha := s + \mu - 4 < 0 , \tag{7}$$

where the negative sign of α gives

$$0 < \sigma := 2 + \alpha - \frac{\mu}{2} < 2 + \frac{\alpha - \mu}{2} =: \sigma' . \tag{8}$$

Note that we may suppose without loss of generality that $|\alpha|$ is sufficiently small in order to obtain the positive sign of σ. Alternatively, we observe that in the case of a negative sign the second integral on the right-hand side of the inequality (13) below is trivially bounded.

By (8) we may choose in addition $k \in \mathbb{N}$ sufficiently large satisfying

$$2k \frac{\sigma}{\sigma'} < 2k - 2 .$$

Now, given $\eta \in C_0^\infty(B)$, $0 \le \eta \le 1$, $\eta \equiv 1$ on $B_r(x_0)$, $|\nabla \eta| \le c/(R - r)$, we introduce the function $\Gamma_\delta = 1 + |\nabla u_\delta|^2$ and recall (6). Hence, u_δ is smooth enough to perform the integration by parts

$$\int_B |\nabla u_\delta|^2 \Gamma_\delta^{1 + \frac{\alpha - \mu}{2}} \eta^{2k} \, dx = -\int_B u_\delta^i \cdot \nabla \left[\nabla u_\delta^i \Gamma_\delta^{1 + \frac{\alpha - \mu}{2}} \eta^{2k} \right] \, dx$$

$$\le c \int_B |\nabla^2 u_\delta| \Gamma_\delta^{1 + \frac{\alpha - \mu}{2}} \eta^{2k} \, dx$$

$$+ c \int_B \Gamma_\delta^{\frac{3 + \alpha - \mu}{2}} \eta^{2k - 1} |\nabla \eta| \, dx .$$

Here we already made use of the fact that u_δ is uniformly bounded. If a positive constant M is fixed, then the left-hand side is immediately estimated by

$$\int_B |\nabla u_\delta|^2 \Gamma_\delta^{1+\frac{\alpha-\mu}{2}} \eta^{2k}\, dx \geq c \int_{B \cap [|\nabla u_\delta| \geq M]} \Gamma_\delta^{2+\frac{\alpha-\mu}{2}} \eta^{2k}\, dx$$

$$\geq c \int_B \Gamma_\delta^{2+\frac{\alpha-\mu}{2}} \eta^{2k}\, dx - c(M)\,,$$

therefore the starting inequality reads as

$$\int_B \Gamma_\delta^{2+\frac{\alpha-\mu}{2}} \eta^{2k}\, dx \leq c\left\{ 1 + \int_B |\nabla^2 u_\delta| \Gamma_\delta^{1+\frac{\alpha-\mu}{2}} \eta^{2k}\, dx \right.$$

$$\left. + \int_B \Gamma_\delta^{\frac{3+\alpha-\mu}{2}} \eta^{2k-1} |\nabla\eta|\, dx \right\} \tag{9}$$

$$=: c\{1 + I + II\}\,.$$

At this point we like to emphasize that the choice (7) of α gives

$$2 + \frac{\alpha-\mu}{2} = \frac{s}{2} > \frac{q}{2}\,. \tag{10}$$

Now for $\varepsilon > 0$ sufficiently small Young's inequality yields a bound for II

$$II \leq \varepsilon \int_B \Gamma_\delta^{2+\frac{\alpha-\mu}{2}} \eta^{2k}\, dx + \varepsilon^{-1} \int_B \Gamma_\delta^{-2-\frac{\alpha-\mu}{2}} \Gamma_\delta^{3+\alpha-\mu} \eta^{2k-2} |\nabla\eta|^2\, dx$$

$$\leq \varepsilon \int_B \Gamma_\delta^{2+\frac{\alpha-\mu}{2}} \eta^{2k}\, dx + \frac{c\varepsilon^{-1}}{(R-r)^2} \int_B \Gamma_\delta^{1+\frac{\alpha-\mu}{2}} \eta^{2k-2}\, dx\,. \tag{11}$$

Note that the first integral on the right-hand side of (11) may be absorbed on the left-hand side of (9), whereas the second one remains uniformly bounded on account of Remark 5.3, $ii)$, the uniform bound of $\int f(\nabla u_\delta)\, dx$ and $\alpha < 0$. Hence, the theorem is proved if an appropriate estimate for I is found. To this purpose we observe (again $\varepsilon > 0$ is sufficiently small and Young's inequality is applied)

$$I \leq \varepsilon \int_B \Gamma_\delta^{-\frac{\mu}{2}} |\nabla^2 u_\delta|^2 \eta^{2k+2}\, dx + \varepsilon^{-1} \int_B \Gamma_\delta^{2+\alpha-\frac{\mu}{2}} \eta^{2k-2}\, dx \tag{12}$$

$$=: \varepsilon I_1 + \varepsilon^{-1} I_2\,.$$

Using Lemma 5.9, $iv)$, as well as Lemma 5.10 one obtains

$$I_1 \leq \int_B D^2 f_\delta(\nabla u_\delta)(\partial_\gamma \nabla u_\delta, \partial_\gamma \nabla u_\delta)\left(\eta^{k+1}\right)^2 dx$$

$$\leq c \int_B \left|D^2 f_\delta(\nabla u_\delta)\right| |\nabla u_\delta|^2 \eta^{2k} |\nabla\eta|^2 dx$$

$$\leq \frac{c}{(R-r)^2}\left\{\int_B \Gamma_\delta^{\frac{3}{2}} \eta^{2k} dx + \delta \int_B \Gamma_\delta^{\frac{t}{2}} \eta^{2k} dx\right\}$$

$$\leq \frac{c}{(R-r)^2}\left\{1 + \int_B \Gamma_\delta^{\frac{3}{2}} \eta^{2k} dx\right\}.$$

As a result, (12) yields (using (10))

$$I \leq \frac{c\varepsilon}{(R-r)^2}\left\{1 + \int_B \Gamma_\delta^{2+\frac{\alpha-\mu}{2}} \eta^{2k} dx\right\} + \varepsilon^{-1} \int_B \Gamma_\delta^{2+\alpha-\frac{\mu}{2}} \eta^{2k-2} dx . \tag{13}$$

Choosing $\varepsilon = \hat{\varepsilon}(R-r)^2$ with $\hat{\varepsilon} > 0$ sufficiently small, the first integral on the right-hand side of (13) may also be absorbed on the left-hand side of (9), hence it remains to bound the second one. Here, the negative sign of α and, as a consequence, (8) and our choice of k come into play. For $\tilde{\varepsilon} > 0$ sufficiently small we get with a final application of Young's inequality

$$\hat{\varepsilon}^{-1}(R-r)^{-2}\int_B \Gamma_\delta^{2+\alpha-\frac{\mu}{2}} \eta^{2k-2} dx$$

$$\leq c\hat{\varepsilon}^{-1}(R-r)^{-2}\left\{\tilde{\varepsilon}\int_B \Gamma_\delta^{2+\frac{\alpha-\mu}{2}} \eta^{2k} dx + \tilde{\varepsilon}^{-\frac{\sigma}{\sigma'-\sigma}} |B|\right\}.$$

Absorbing terms by letting $\tilde{\varepsilon} = \varepsilon' \hat{\varepsilon}(R-r)^2$, $1 \gg \varepsilon' > 0$, Theorem 5.12 is proved implying the validity of Theorem 5.4 as well. □

5.2 Scalar obstacle problems

Let us now turn our attention to scalar problems, where we already mentioned in the introduction that now – as a model case – obstacle problems are included whereas we omit the discussion of vector-valued problems with additional structure. The general hypothesis under consideration is slightly different from Assumption 5.1:

Assumption 5.13. *Let $N = 1$ and assume that the energy density $f \colon \mathbb{R}^n \to [0,\infty)$ satisfies $f(0) = 0$ and $\nabla f(0) = 0$. Moreover, suppose that f is a strictly convex function of class $C^2(\mathbb{R}^n)$ which satisfies (1) and the relation (4) of the previous section. Condition (3), Section 5.1, now merely is assumed for all $|Z| > 1$.*

Remark 5.14.

i) *Clearly this setting is much more general than the one considered in [Ch]:
we do not suppose $f(Z) = g(|Z|^2)$ and we just assume the relation $q <
4 - \mu$ instead of $q < p + 1$.
As a formal difference, Choe studies energy densities admitting some kind
of degeneracy as $|Z| \to 0$. This behavior of the second derivative is covered
by Assumption 5.13 since (3), Section 5.1, is not supposed in the case
$|Z| < 1$. We already like to remark that this causes no additional technical
difficulties since in any way we have to rely on a suitable cut-off function
in order to study obstacle problems.*

ii) *Further comments on Assumption 5.13 were already made in Remark 5.3.*

The second main result of this chapter is made precise in

Theorem 5.15. *Fix boundary data u_0 according to (2), Section 5.1. Moreover, let the energy density f satisfy Assumption 5.13 and suppose that we are
given a function ψ of class $W_{\infty,loc}^1(\Omega)$ such that $u_0 \geq \psi$ almost everywhere.
Finally, let*

$$u \in \mathbb{K}_F^\psi := \mathbb{K}_F \cap \left\{ w \in \mathbb{K}_F : w \geq \psi \text{ almost everywhere} \right\}$$

*denote the uniquely determined minimizer of the problem (P) with respect to
comparison functions of class $\mathbb{K}_F^\psi$. Then we have:*

i) *u is of class $W_{\infty,loc}^1(\Omega)$.*

ii) *If in addition the inequality (3), Section 5.1, holds for any $Z \in \mathbb{R}^n$ then
u is of class $C^{1,\alpha}(\Omega)$, if so is the obstacle.*

Remark 5.16.

i) *The existence and uniqueness of solutions to scalar obstacle problems follows exactly in the same manner as outlined in Section 3.1.*

ii) *A discussion of (maybe degenerate) obstacle problems is found, for instance, in the papers [MiZ], [CL], [Lin], [MuZ], [Fu1] and [Fu2]. The classical quadratic case is extensively treated in the monographs [KS], [Fri].
Non-standard growth conditions are considered in [Lie1], [BFM].*

iii) *If we consider degenerate energy densities with (p, q)-growth, then local
Lipschitz continuity is improved to $C_{loc}^{1,\alpha}$-regularity following [BFM] (compare [MuZ]). Here an additional hypothesis is needed to control the kind
of degeneration of $D^2 f$.*

iv) *Remark 5.5 remains unchanged. In particular from ii), Remark 5.5, we
see the substantial improvement of admissible exponents in the case that
u_0 satisfies (2), Section 5.1 (additionally recall the comparison with known
results given in Section 3.5).*

v) Note that our assumptions are strong enough to obtain a L^∞-bound for u. In fact, we take $w := \min\{u, \sup_\Omega u_0\}$ as a comparison function and recall that $f(0) = 0$ and $\nabla f(0) = 0$. Hence, the strictly convex integrand f attains its minimum in 0 and we get $u \leq \sup_\Omega u_0$ (an analogous estimate holds for the negative sign). Moreover, the same arguments are valid for the regularization introduced below.

The *Proof of Theorem 5.15* splits into four steps: regularization, linearization and Caccioppoli-type inequalities, uniform local higher integrability and uniform local a priori gradient bounds, where we always assume that the hypotheses of Theorem 5.15 are satisfied.

Step 1. (Regularization) Choose $(u)_\varepsilon$ and B as in the proof of Theorem 5.4 and let $(\psi)_\varepsilon$ denote the ε-mollification of the obstacle. Then u_δ $(= u_\delta^\varepsilon)$ is defined as the unique solution of the problem

$$J_\delta[w, B] := \int_B f_\delta(\nabla w) \, dx \to \min , \quad w \in \mathbb{K}_\varepsilon^\psi , \qquad (\mathcal{P}_\delta^{(\psi)_\varepsilon})$$

where $\mathbb{K}_\varepsilon^\psi = \{w \in (u)_{|B}^\varepsilon + \overset{\circ}{W}_q^1(B) : (\psi)_\varepsilon \leq w \text{ almost everywhere}\}$ and where we have set

$$f_\delta(Z) := f(Z) + \delta \left(1 + |Z|^2\right)^{\frac{q}{2}} , Z \in \mathbb{R}^n .$$

Remark 5.17. *(compare Remark 3.13 and Remark 5.8) As in Chapter 3 the regularization is done with respect to the exponent q. This essentially reduces the difficulties in deriving uniform local a priori gradient bounds. Note that the problem of starting integrability disappears in the scalar case (see Lemma 5.19 below).*

Remark 5.18. *Lemma 5.9 remains valid in the case of obstacle problems, where we just have to observe: the weak W_1^1- and the almost everywhere convergence of the sequence $\{u_\delta\}$ is proved as before. The limit $\tilde{u}$ respects the obstacle by the almost everywhere convergence, hence lower semicontinuity and the uniqueness of solutions give $\tilde{u} = u$. The remaining assertions follow as above.*

Step 2. (Linearization and Caccioppoli-type inequalities) The study of obstacle problems needs an additional linearization. This procedure is well known, a detailed proof of the following lemma is given in [Ful].

Lemma 5.19. *With the above notation, u_δ is of class $W_{t,loc}^2(B)$ for any $t < \infty$ and*

$$\nabla f_\delta(\nabla u_\delta) \in W_{t,loc}^1(B).$$

Moreover, the equation

$$\int_B \nabla f_\delta(\nabla u_\delta) \cdot \nabla \varphi \, dx = \int_B \varphi g \, dx \qquad (14)$$

is valid for any $\varphi \in C_0^1(B)$, where

$$g := \mathbf{1}_{\{x \in B:\, u_\delta = (\psi)_\varepsilon\}} \left(- \operatorname{div} \left[\nabla f_\delta (\nabla(\psi)_\varepsilon) \right] \right) .$$

Given Lemma 5.19, we now formulate two Caccioppoli-type inequalities on level-sets.

Lemma 5.20. *Suppose that the hypotheses of Theorem 5.15 are satisfied and fix $L > 1$ such that for any ε as above*

$$L > 1 + \|\nabla(\psi)_\varepsilon\|_{L^\infty(B;\mathbb{R}^n)}^2 .$$

i) Let $B_\varkappa := \{x \in B : \Gamma_\delta := 1 + |\nabla u_\delta|^2 > \varkappa\}$, $\varkappa > 1$. Then there is a constant c, independent of δ (and ε), such that for any $\varkappa > L$, for any real number $s \geq 0$ and for any $\eta \in C_0^\infty(B)$, $0 \leq \eta \leq 1$,

$$\int_{B_{2\varkappa}} D^2 f_\delta(\nabla u_\delta)(\partial_\gamma \nabla u_\delta, \partial_\gamma \nabla u_\delta) \Gamma_\delta^s \eta^2 \, dx$$

$$\leq c \int_{B_\varkappa} \left| D^2 f_\delta(\nabla u_\delta) \right| \Gamma_\delta^{1+s} |\nabla \eta|^2 \, dx .$$

ii) Recall that $\Gamma_\delta := 1 + |\nabla u_\delta|^2$ and let for $0 < r < R$

$$A(k,r) = A_\delta(k,r) = \{x \in B_r(x_0) : \Gamma_\delta > k\}, \quad k > 1 + L .$$

Then there is a real number $c > 0$ such that for any $\eta \in C_0^\infty(B_r(x_0))$, $0 \leq \eta \leq 1$ and for any $\delta \in (0,1)$

$$\int_{A(k,r)} \Gamma_\delta^{-\frac{\mu}{2}} |\nabla \Gamma_\delta|^2 \eta^2 \, dx$$

$$\leq c \int_{A(k,r)} D^2 f_\delta(\nabla u_\delta)(\nabla \eta, \nabla \eta) (\Gamma_\delta - k)^2 \, dx .$$

Proof. ad i). This time we shortly sketch the proof following the idea given in [BFM], Lemma 2.3: fix $\varkappa > L$ and let for all $t \in \mathbb{R}$

$$\tilde{h}(t) := \min \left\{ \max[t - 1, 0], 1 \right\}, \quad h(t) = h_\varkappa(t) = \tilde{h}(\varkappa^{-1}t), \tag{15}$$

i.e. $h(t) \equiv 0$ if $t < \varkappa$ and $h(t) \equiv 1$ if $t > 2\varkappa$. The integrability result of Lemma 5.19 allows us to differentiate the equation (14), thus

$$\int_B D^2 f_\delta(\nabla u_\delta) \left(\partial_\gamma \nabla u_\delta, \nabla \left(\eta^2 \, \partial_\gamma u_\delta \, h(\Gamma_\delta) \Gamma_\delta^s \right) \right) dx$$

$$= - \int_B g \partial_\gamma \left(\eta^2 \, \partial_\gamma u_\delta \, h(\Gamma_\delta) \Gamma_\delta^s \right) dx .$$

On the set of coincidence we have almost everywhere $\nabla u_\delta = \nabla(\psi)_\varepsilon$ (see [GT], Lemma 7.7, p. 152), hence the auxiliary function $h(\Gamma_\delta)$ vanishes on this set by the choice $\varkappa > L$. This, together with

$$\int_B D^2 f_\delta(\nabla u_\delta)(\partial_\gamma u_\delta \, \partial_\gamma \nabla u_\delta, \nabla \Gamma_\delta) \, h'(\Gamma_\delta) \Gamma_\delta^s \eta^2 \, dx \geq 0 ,$$

$$s \int_B D^2 f_\delta(\nabla u_\delta)(\partial_\gamma u_\delta \, \partial_\gamma \nabla u_\delta, \nabla \Gamma_\delta) \Gamma_\delta^{s-1} h(\Gamma_\delta) \eta^2 \, dx \geq 0$$

(which follows from $h' \geq 0$, $s \geq 0$ and $2\partial_\gamma u_\delta \, \partial_\gamma \nabla u_\delta = \nabla \Gamma_\delta$) yields

$$\int_B D^2 f_\delta(\nabla u_\delta)(\partial_\gamma \nabla u_\delta, \partial_\gamma \nabla u_\delta) \, h(\Gamma_\delta) \Gamma_\delta^s \eta^2 \, dx$$

$$\leq - \int_B D^2 f_\delta(\nabla u_\delta)(\partial_\gamma \nabla u_\delta, \nabla \eta) \, \eta \, \partial_\gamma u_\delta \, h(\Gamma_\delta) \Gamma_\delta^s \, dx .$$

Finally, Young's inequality proves the claim after absorbing terms.

ad ii). Following the reasoning of Lemma 4.19, ii), we now have to include the side condition. If we are given $k > 1 + L$ then we choose $\varphi = \eta^2 \, \partial_\gamma u_\delta \max \left[\Gamma_\delta - k, 0 \right]$, η as above. Again Lemma 5.19 shows the validity of the equation (14) and its differentiated version. As before the right-hand side vanishes since k is large enough, thus

$$\int_{A(k,r)} D^2 f_\delta(\nabla u_\delta)(\partial_\gamma \nabla u_\delta, \partial_\gamma \nabla u_\delta)(\Gamma_\delta - k) \eta^2 \, dx$$

$$+ \int_{A(k,r)} D^2 f_\delta(\nabla u_\delta)(\partial_\gamma \nabla u_\delta, \nabla \Gamma_\delta) \, \partial_\gamma u_\delta \, \eta^2 \, dx \tag{16}$$

$$= -2 \int_{A(k,r)} D^2 f_\delta(\nabla u_\delta)(\partial_\gamma \nabla u_\delta, \nabla \eta) \, \eta \, \partial_\gamma u_\delta \, (\Gamma_\delta - k) \, dx .$$

Here the non-negative first integral on the left-hand side is neglected, the second one satisfies:

$$\int_{A(k,r)} D^2 f_\delta(\nabla u_\delta)(\partial_\gamma \nabla u_\delta, \nabla \Gamma_\delta) \, \partial_\gamma u_\delta \, \eta^2 \, dx$$

$$= \frac{1}{2} \int_{A(k,r)} D^2 f_\delta(\nabla u_\delta)(\nabla \Gamma_\delta, \nabla \Gamma_\delta) \, \eta^2 \, dx . \tag{17}$$

The right-hand side of (16) is estimated from above by

$$c\varepsilon \int_{A(k,r)} D^2 f_\delta(\nabla u_\delta)(\nabla \Gamma_\delta, \nabla \Gamma_\delta) \, \eta^2 \, dx$$

$$+ c\varepsilon^{-1} \int_{A(k,r)} D^2 f_\delta(\nabla u_\delta)(\nabla \eta, \nabla \eta)(\Gamma_\delta - k)^2 \, dx , \tag{18}$$

where we made use of Young's inequality for $\varepsilon > 0$ sufficiently small. Absorbing terms the lemma is proved by (16)–(18) and the ellipticity condition (3), Section 5.1, which can be applied on account of $k > 1 + L$. □

Step 3. (Uniform local higher integrability) In the case of scalar obstacle problems higher integrability is improved by an iteration argument to

Theorem 5.21. *Recall our general assumptions in the case $N = 1$. Then for any $1 < s < \infty$ and for any ball $B_r(x_0)$, $r < R$, there is a constant c, just depending on the data, $\sup_B |(u)_\varepsilon|$, r and s, such that*

$$\int_{B_r(x_0)} |\nabla u_\delta|^s \, dx \leq c < \infty .$$

Proof. We now fix some non-negative number $\alpha \geq 0$ and let

$$\beta = 4 - \mu - q > 0 ,$$

where the positive sign follows from the assumption (4), Section 5.1. As a consequence, the counterpart of (8), Section 5.1, reads as

$$0 < \sigma := 2 + \frac{\alpha - \beta - \mu}{2} < 2 + \frac{\alpha - \mu}{2} =: \sigma' .$$

Again we may choose $k \in \mathbb{N}$ sufficiently large such that

$$2k \frac{\sigma}{\sigma'} < 2k - 2 .$$

Next we have to make use of the auxiliary function h defined in (15). Here we define h with respect to $2\varkappa$, $\varkappa > L + 1$, L given in Lemma 5.20. Once more, the starting inequality is derived by performing an integration by parts which is admissible on account of Lemma 5.19

$$\int_B |\nabla u_\delta|^2 \Gamma_\delta^{1+\frac{\alpha-\mu}{2}} h(\Gamma_\delta)\eta^{2k} \, dx = - \int_B u_\delta \partial_\gamma \left[\partial_\gamma u_\delta \Gamma_\delta^{1+\frac{\alpha-\mu}{2}} h(\Gamma_\delta)\eta^{2k} \right] dx$$

$$\leq c \int_B |\nabla^2 u_\delta| \Gamma_\delta^{1+\frac{\alpha-\mu}{2}} h(\Gamma_\delta)\eta^{2k} \, dx$$

$$+ c \int_B \Gamma_\delta^{\frac{3+\alpha-\mu}{2}} h(\Gamma_\delta)\eta^{2k-1} |\nabla\eta| \, dx$$

$$+ c \int_B \Gamma_\delta^{\frac{3+\alpha-\mu}{2}} h'(\Gamma_\delta)|\nabla u_\delta||\nabla^2 u_\delta|\eta^{2k} \, dx .$$

This time it is supposed that $r < \rho < \rho' \leq R$, $\eta \in C_0^\infty(B_{\rho'}(x_0))$, $\eta \equiv 1$ on $B_\rho(x_0)$, $|\nabla\eta| \leq c(\rho' - \rho)^{-1}$. A lower bound for the left-hand side of this inequality is given by (compare Section 5.1)

$$\int_B \Gamma_\delta^{2+\frac{\alpha-\mu}{2}} \eta^{2k} \, dx - c(\varkappa) ,$$

on the right-hand side we observe that $h'(\Gamma_\delta)$ identically vanishes outside the set $[2\varkappa \leq \Gamma_\delta \leq 4\varkappa]$. As an immediate consequence we have

$$\int_B \Gamma_\delta^{\frac{3+\alpha-\mu}{2}} \, h'(\Gamma_\delta) |\nabla u_\delta| |\nabla^2 u_\delta| \eta^{2k} \, dx \leq c(\varkappa) \int_{B_{2\varkappa}} |\nabla^2 u_\delta| \Gamma_\delta^{1+\frac{\alpha-\mu}{2}} \eta^{2k} \, dx \, ,$$

where the definition of $B_{2\varkappa}$ is the same as in Lemma 5.20. Since it is also obvious that

$$\int_B |\nabla^2 u_\delta| \Gamma_\delta^{1+\frac{\alpha-\mu}{2}} \, h(\Gamma_\delta) \eta^{2k} \, dx \leq \int_{B_{2\varkappa}} |\nabla^2 u_\delta| \Gamma_\delta^{1+\frac{\alpha-\mu}{2}} \eta^{2k} \, dx \, ,$$

and since an analogous estimate holds for the remaining integral, we arrive at

$$\int_B \Gamma_\delta^{2+\frac{\alpha-\mu}{2}} \eta^{2k} \, dx \leq c \bigg\{ 1 + \int_{B_{2\varkappa}} |\nabla^2 u_\delta| \Gamma_\delta^{1+\frac{\alpha-\mu}{2}} \eta^{2k} \, dx$$

$$+ \int_{B_{2\varkappa}} \Gamma_\delta^{\frac{3+\alpha-\mu}{2}} \eta^{2k-1} |\nabla \eta| \, dx \bigg\} \tag{19}$$

$$=: c\{1 + I + II\} \, .$$

Now, given $\varepsilon > 0$ sufficiently small, II is handled in the same manner as in Section 5.1

$$II \leq \varepsilon \int_B \Gamma_\delta^{2+\frac{\alpha-\mu}{2}} \eta^{2k} \, dx + \frac{c\varepsilon^{-1}}{(\rho' - \rho)^2} \int_B \Gamma_\delta^{1+\frac{\alpha-\mu}{2}} \eta^{2k-2} \, dx \, , \tag{20}$$

where the first integral can be absorbed on the left-hand side of (19). For the discussion of I we first observe that

$$I \leq \varepsilon \int_{B_{2\varkappa}} \Gamma_\delta^{\frac{\alpha+\beta}{2}} \Gamma_\delta^{-\frac{\mu}{2}} |\nabla^2 u_\delta|^2 \eta^{2k+2} \, dx$$

$$+ \varepsilon^{-1} \int_{B_{2\varkappa}} \Gamma_\delta^{2+\frac{\alpha-\beta-\mu}{2}} \eta^{2k-2} \, dx =: \varepsilon I_1 + \varepsilon^{-1} I_2 \, .$$

Here we have to check that I_1 can be handled with the help of Lemma 5.20, i): by definition it is clear that $\alpha + \beta \geq 0$. Moreover, the choice of $\varkappa$ shows that we have (3), Section 5.1, on the set $B_\varkappa$ (recall that (3), Section 5.1, now merely is assumed whenever $|Z| > 1$). One gets

$$I_1 \leq c \int_{B_{2\varkappa}} D^2 f_\delta(\nabla u_\delta)(\partial_\gamma \nabla u_\delta, \partial_\gamma \nabla u_\delta) \Gamma_\delta^{\frac{\alpha+\beta}{2}} \left(\eta^{k+1}\right)^2 \, dx$$

$$\leq c \int_{B_\varkappa} |D^2 f_\delta(\nabla u_\delta)| \Gamma_\delta^{1+\frac{\alpha+\beta}{2}} \eta^{2k} |\nabla \eta|^2 \, dx$$

$$\leq \frac{c}{(\rho' - \rho)^2} \int_B \Gamma_\delta^{\frac{\alpha+\beta}{2}} \Gamma_\delta^{\frac{q}{2}} \eta^{2k} \, dx \, .$$

For the last inequality we recall that the regularization was done with respect to $t = q$. Finally, the choice of β implies

$$I \leq \frac{c\varepsilon}{(\rho' - \rho)^2} \int_B \Gamma_\delta^{2 + \frac{\alpha - \mu}{2}} \eta^{2k} \, dx + \varepsilon^{-1} \int_B \Gamma_\delta^{2 + \frac{\alpha - \beta - \mu}{2}} \eta^{2k-2} \, dx . \qquad (21)$$

If again $\varepsilon = \hat{\varepsilon}(\rho' - \rho)^2$ and if $\hat{\varepsilon} > 0$ is sufficiently small, then we argue exactly as in Section 5.1, i.e. the first integral on the right-hand side of (21) is absorbed on the left-hand side of (19), whereas

$$\hat{\varepsilon}^{-1} (\rho' - \rho)^{-2} \int_B \Gamma_\delta^{2 + \frac{\alpha - \beta - \mu}{2}} \eta^{2k-2} \, dx$$
$$\leq c\hat{\varepsilon}^{-1} (\rho' - \rho)^{-2} \left\{ \tilde{\varepsilon} \int_B \Gamma_\delta^{2 + \frac{\alpha - \mu}{2}} \eta^{2k} \, dx + \tilde{\varepsilon}^{-\frac{\sigma}{\sigma' - \sigma}} |B| \right\} . \qquad (22)$$

Following (19)–(22), letting $\tilde{\varepsilon} = \varepsilon' \hat{\varepsilon}(\rho' - \rho)^2$, $1 \gg \varepsilon' > 0$ and absorbing terms for a last time we have found a real number $c = c(\varkappa, \alpha, \rho' - \rho, \sup_B |(u)_\varepsilon|)$, independent of δ, such that

$$\int_B \Gamma_\delta^{2 + \frac{\alpha - \mu}{2}} \eta^{2k} \, dx \leq c \left\{ 1 + \int_B \Gamma_\delta^{1 + \frac{\alpha - \mu}{2}} \eta^{2k-2} \, dx \right\} . \qquad (23)$$

To start an iteration of (23) let

$$\rho_m = r + (R - r) 2^{-m} , \qquad m = 0, 1, 2, \dots ,$$

as well as

$$\alpha_m = 2m , \quad \text{i.e.} \quad \alpha_{m+1} = 2 + \alpha_m , \qquad m = 0, 1, 2, \dots ,$$

where for any m as above α_m is non-negative, hence admissible in the above calculations. Then we obtain (23) for any $m = 0, 1, 2, \dots$, with the choices $\rho = \rho_{m+1}$, $\rho' = \rho_m$, $\alpha = \alpha_m$, i.e.

$$\int_{B_{\rho_{m+1}}(x_0)} \Gamma_\delta^{1 + \frac{\alpha_{m+1} - \mu}{2}} \, dx \leq c \left\{ 1 + \int_{B_{\rho_m}(x_0)} \Gamma_\delta^{1 + \frac{\alpha_m - \mu}{2}} \, dx \right\} .$$

An iteration completes the proof since $\alpha_0 = 0$ gives a uniformly bounded right-hand side (once more compare Remark 5.3, ii)). □

Step 4. (Uniform local a priori gradient bounds) With Theorem 5.21 we can apply a Moser-type iteration (as done in [Ch]) to obtain uniform local a priori gradient bounds. We prefer the modification of DeGiorgi's technique (similar to Section 4.2.4) which seems to be more convenient in the setting of "bad" ellipticity, moreover, the side condition is easily eliminated.

Theorem 5.22. *Consider a ball $B_{R_0}(x_0) \Subset B$ where we again suppose that the assumptions of Theorem 5.15 are satisfied. Then there is a local constant $c > 0$ such that for any $\delta \in (0, 1)$*

$$\|\nabla u_\delta\|_{L^\infty(B_{R_0/2}, \mathbb{R}^n)} \leq c .$$

Before proving Theorem 5.22 we establish the generalized version of Lemma 4.29, where the case $q = 2$ was under consideration.

Lemma 5.23. *Suppose $0 < r < \hat{r} < R_0$ such that $B_{R_0}(x_0) \Subset B$. Then there is a real number c, independent of r, $\hat{r}$, R_0, k and δ, satisfying for any $k > 1 + L$ (L as above)*

$$\int_{A(k,r)} (\Gamma_\delta - k)^{\frac{n}{n-1}} \, dx$$

$$\leq \frac{c}{(\hat{r} - r)^{\frac{n}{n-1}}} \left[\int_{A(k,\hat{r})} \Gamma_\delta^{\frac{q-2}{2}} (\Gamma_\delta - k)^2 \, dx \right]^{\frac{1}{2}\frac{n}{n-1}} \left[\int_{A(k,\hat{r})} \Gamma_\delta^{\frac{\mu}{2}} \, dx \right]^{\frac{1}{2}\frac{n}{n-1}}, \tag{24}$$

where the sets $A(k,r) = \{x \in B_r(x_0) : \Gamma_\delta > k\}$ are introduced in Lemma 5.20.

Proof of Lemma 5.23. Given $\eta \in C_0^\infty(B_{\hat{r}}(x_0))$, $0 \leq \eta \leq 1$, $\eta \equiv 1$ on $B_r(x_0)$, $|\nabla \eta| \leq c/(\hat{r} - r)$, we proceed exactly as in Lemma 4.29, hence

$$\int_{A(k,r)} (\Gamma_\delta - k)^{\frac{n}{n-1}} \, dx \leq c \left[I_1^{\frac{n}{n-1}} + I_2^{\frac{n}{n-1}} \right],$$

where now the first integral on the right-hand side is estimated from above by (recall $2 - \mu \leq q$)

$$I_1^{\frac{n}{n-1}} = \left[\int_{A(k,\hat{r})} |\nabla \eta| (\Gamma_\delta - k) \, dx \right]^{\frac{n}{n-1}}$$

$$\leq \left[\int_{A(k,\hat{r})} |\nabla \eta|^2 \Gamma_\delta^{\frac{q-2}{2}} (\Gamma_\delta - k)^2 \, dx \right]^{\frac{1}{2}\frac{n}{n-1}} \left[\int_{A(k,\hat{r})} \Gamma_\delta^{\frac{2-q}{2}} \, dx \right]^{\frac{1}{2}\frac{n}{n-1}}$$

$$\leq \frac{c}{(\hat{r} - r)^{\frac{n}{n-1}}} \left[\int_{A(k,\hat{r})} \Gamma_\delta^{\frac{q-2}{2}} (\Gamma_\delta - k)^2 \, dx \right]^{\frac{1}{2}\frac{n}{n-1}} \left[\int_{A(k,\hat{r})} \Gamma_\delta^{\frac{\mu}{2}} \, dx \right]^{\frac{1}{2}\frac{n}{n-1}}.$$

Discussing I_2 we recall the choice $k > 1 + L$. Thus it is possible to refer to Lemma 5.20, *ii*), with the result

$$I_2^{\frac{n}{n-1}} = \left[\int_{A(k,\hat{r})} \eta |\nabla \Gamma_\delta| \, dx \right]^{\frac{n}{n-1}}$$

$$\leq \left[\int_{A(k,\hat{r})} \eta^2 |\nabla \Gamma_\delta|^2 \Gamma_\delta^{-\frac{\mu}{2}} \, dx \right]^{\frac{1}{2}\frac{n}{n-1}} \left[\int_{A(k,\hat{r})} \Gamma_\delta^{\frac{\mu}{2}} \, dx \right]^{\frac{1}{2}\frac{n}{n-1}}$$

$$\leq c \left[\int_{A(k,\hat{r})} D^2 f_\delta(\nabla u_\delta)(\nabla \eta, \nabla \eta) (\Gamma_\delta - k)^2 \, dx \right]^{\frac{1}{2}\frac{n}{n-1}} \left[\int_{A(k,\hat{r})} \Gamma_\delta^{\frac{\mu}{2}} \, dx \right]^{\frac{1}{2}\frac{n}{n-1}}$$

$$\leq \frac{c}{(\hat{r} - r)^{\frac{n}{n-1}}} \left[\int_{A(k,\hat{r})} \Gamma_\delta^{\frac{q-2}{2}} (\Gamma_\delta - k)^2 \, dx \right]^{\frac{1}{2}\frac{n}{n-1}} \left[\int_{A(k,\hat{r})} \Gamma_\delta^{\frac{\mu}{2}} \, dx \right]^{\frac{1}{2}\frac{n}{n-1}}$$

and the lemma is proved. $\square$

Proof of Theorem 5.22. Again we merely have to modify the reasoning of Section 4.2.4. Starting with the left-hand side of (24), we fix a real number $s > 1$ and observe that Hölder's inequality implies

$$\int_{A(k,r)} \Gamma_\delta^{\frac{q-2}{2}} (\Gamma_\delta - k)^2 \, dx = \int_{A(k,r)} (\Gamma_\delta - k)^{\frac{n}{n-1}\frac{1}{s}} \Gamma_\delta^{\frac{q-2}{2}} (\Gamma_\delta - k)^{2-\frac{n}{n-1}\frac{1}{s}} \, dx$$

$$\leq \left[\int_{A(k,r)} (\Gamma_\delta - k)^{\frac{n}{n-1}} \, dx \right]^{\frac{1}{s}}$$

$$\cdot \left[\int_{A(k,r)} \Gamma_\delta^{\frac{q-2}{2}\frac{s}{s-1}} (\Gamma_\delta - k)^{\left(2-\frac{n}{n-1}\frac{1}{s}\right)\frac{s}{s-1}} \right]^{\frac{s-1}{s}}.$$

Theorem 5.21 gives a real number $c_1(s,n,B_{R_0}(x_0))$, independent of δ,

$$c_1(s,n,B_{R_0}(x_0)) := \sup_{\delta>0} \left[\int_{B_{R_0}(x_0)} \Gamma_\delta^{\frac{s}{s-1}\left(\frac{q-2}{2}+2-\frac{n}{n-1}\frac{1}{s}\right)} \, dx \right]^{\frac{s-1}{s}} < \infty,$$

such that

$$\int_{A(k,r)} \Gamma_\delta^{\frac{q-2}{2}} (\Gamma_\delta - k)^2 \, dx \leq c_1 \left[\int_{A(k,r)} (\Gamma_\delta - k)^{\frac{n}{n-1}} \, dx \right]^{\frac{1}{s}}. \qquad (25)$$

In a similar way one obtains

$$\int_{A(k,\hat{r})} \Gamma_\delta^{\frac{\mu}{2}} \, dx \leq c_2(t,\mu,B_{R_0}(x_0)) \left[\int_{A(k,\hat{r})} \Gamma_\delta^{\frac{q-2}{2}} \, dx \right]^{\frac{1}{t}}, \qquad (26)$$

where $t > 1$ is a fixed second parameter. Combining (24), (25) and (26) it is proved that

$$\int_{A(k,r)} \Gamma_\delta^{\frac{q-2}{2}} (\Gamma_\delta - k)^2 \, dx \leq \frac{c}{(\hat{r}-r)^{\frac{n}{n-1}\frac{1}{s}}} \left[\int_{A(k,\hat{r})} \Gamma_\delta^{\frac{q-2}{2}} (\Gamma_\delta - k)^2 \, dx \right]^{\frac{1}{2}\frac{n}{n-1}\frac{1}{s}}$$

$$\cdot \left[\int_{A(k,\hat{r})} \Gamma_\delta^{\frac{q-2}{2}} \, dx \right]^{\frac{1}{2}\frac{n}{n-1}\frac{1}{s}\frac{1}{t}}. \qquad (27)$$

For $k > 1 + L$ and for $0 < r < \hat{r}$ as above let

$$\tau(k,r) := \int_{A(k,r)} \Gamma_\delta^{\frac{q-2}{2}} (\Gamma_\delta - k)^2 \, dx, \qquad a(k,r) := \int_{A(k,r)} \Gamma_\delta^{\frac{q-2}{2}} \, dx.$$

As before we obtain $(h > k > 1 + L)$

$$\tau(h,r) \leq \frac{c}{(\hat{r}-r)^{\frac{n}{n-1}\frac{1}{s}}} \frac{1}{(h-k)^{\frac{n}{n-1}\frac{1}{s}\frac{1}{t}}} \left[\tau(k,\hat{r}) \right]^{\frac{1}{2}\frac{n}{n-1}\frac{1}{s}\left(1+\frac{1}{t}\right)}$$

and the choice of parameters given in Section 4.2.4 completes the proof. $\square$

Since the data of the obstacle just enter through the constant L, the *Proof of Theorem 5.15, i),* is an immediate consequence of Lemma 5.9 (compare Remark 5.18). Now that $i)$ is established, the second claim follows from the well known papers [MuZ] and [CL] (compare also [FuM] for details).   □

Remark 5.24. *As outlined in the introduction of this chapter, the techniques of Section 5.1 and Section 5.2 provide a suitable approach to superlinear problems.*

In the case of variational problems with linear growth conditions, we preferred arguments based on the discussion of additional comparison functions. One reason to use different techniques is found in Lemma 4.33: the proof crucially relies on the absence of second derivatives of u_δ which is achieved by using a test-function independent of ∇u_δ. Starting with an integration by parts (as done in Section 5) we would automatically obtain terms involving $\nabla^2 u_\delta$ making the arguments impossible.

Nevertheless, the close relationship between linear and superlinear growth problems will become evident in the next Chapter where we will use an appropriate version of Theorem 5.22.

6

Anisotropic linear/superlinear growth in the scalar case

Throughout our studies we changed the point of view several times, where the main line skipped from linear to superlinear growth problems and vice versa. In doing so we could benefit from corresponding examples and techniques. For instance, the study of integrands with (s, μ, q)-growth automatically led to the discussion of energy densities with linear growth and μ-ellipticity. Here a boundedness condition substantially improved the results and the same was shown in the following chapter for superlinear growth problems.

Next we propose the study of variational integrands with anisotropic linear/superlinear growth, i.e. of energy densities with anisotropic "plastic/elastic" behavior. This closes the line from Chapter 5 to Chapter 2: on one hand, problems of this type are now within reach since regularity results in the case of variational integrals with a wide range of anisotropy are available by the discussion of Chapter 5. On the other hand, we need a refined study of some elements from duality theory as introduced in Chapter 2.

Before going into details let us recall Example 5.7, *iii*), where we want to restrict to the scalar case $N = 1$ in this chapter. Writing $Z = (Z_1, Z_2) \in \mathbb{R}^k \times \mathbb{R}^{n-k}$ for some $1 \leq k < n$, we consider the energy density

$$f(Z) = \left(1 + |Z|^2\right)^{\frac{p}{2}} + \left(1 + |Z_2|^2\right)^{\frac{q}{2}}$$

with exponents $1 < p < q < 2$. In this subquadratic situation the estimate

$$\lambda \left(1 + |Z|^2\right)^{\frac{p-2}{2}} |Y|^2 \leq D^2 f(Z)(Y, Y) \leq \Lambda |Y|^2$$

for all $Z, Y \in \mathbb{R}^n$ and with positive constants λ, Λ is the best possible one. Hence, there is a quite large difference (independent of $q - p$) of the upper and the lower growth rate if p is close to one. This is the reason why the example does not fit into the discussion of Section 3.3 (recall Section 3.5 for a comparison with other results).

On the other hand, with the results of the previous chapter the case

$$\lambda \left(1 + |Z|^2\right)^{-\frac{\mu}{2}} |Y|^2 \leq D^2 f(Z)(Y, Y) \leq \Lambda \left(1 + |Z|^2\right)^{\frac{q-2}{2}} |Y|^2 \qquad (1)$$

can be handled (if u_0 is of class $L^\infty(\Omega)$) whenever $1 < q < 4 - \mu$ and f is bounded from below by some N-function. Thus, the idea of this chapter is based on a simple observation: if we take the upper exponent $q = 2$ in (1), then ellipticity up to $\mu < 2$ is admissible. Moreover, in the case $1 < \mu < 2$ we have linear growth examples as discussed in Chapter 4 and there is the hope to include anisotropic linear/superlinear growth energy densities in our considerations – at least up to a certain extent. As an interesting application one may think of anisotropic materials somewhere between plastic and elastic behavior.

Of course there will be no one-to-one correspondence to the results of Chapter 5 which is due to the missing existence of solutions in the setting considered here. Let us already mention that some of our main examples even do not admit a dual solution – this is caused by an additional anisotropic behavior of the superlinear part (compare Remark 6.6 and Remark 6.9 below). Nevertheless, Theorem 5.22 is strong enough to imply local smoothness and uniqueness (up to a constant) of generalized minimizers.

The hypothesis of this chapter reads as

Assumption 6.1. *The energy density $f\colon \mathbb{R}^n \to \mathbb{R}$ is of class $C^2(\mathbb{R}^n)$ and admits the decomposition*

$$f(Z) = f_1(Z_1) + f_2(Z_2)\,, \qquad Z = (Z_1, Z_2) \in \mathbb{R}^k \times \mathbb{R}^{n-k}\,, \qquad (2)$$

for some $k \in \mathbb{N}$, $1 \le k < n$. Here $f_1 \in C^2(\mathbb{R}^k)$ is a function of linear growth such that

$$|\nabla f_1(Z_1)| \le A \qquad (3)$$

holds for any $Z_1 \in \mathbb{R}^k$ with some constant A. The function $f_2 \in C^2(\mathbb{R}^{n-k})$ is supposed to satisfy for some $1 < p \le 2$

$$c_1 |Z_2|^p - c_2 \le f_2(Z_2)\,, \qquad (4)$$

now for any $Z_2 \in \mathbb{R}^{n-k}$ and with constants c_1, c_2. Our assumption on the second derivative of $f = f_1 + f_2$ is given by

$$\lambda \left(1 + |Z|^2\right)^{-\frac{\mu}{2}} |Y|^2 \le D^2 f(Z)(Y,Y) \le \Lambda |Y|^2 \qquad (5)$$

for all Z, $Y \in \mathbb{R}^n$ with constants λ, Λ and with an exponent of ellipticity

$$1 < \mu < 2\,. \qquad (6)$$

Remark 6.2. *Let us recall some consequences (compare Remark 3.5, iv), and Remark 4.2 i), ii)).*

i) *Whenever it is needed, we will assume without loss of generality that $f_i(0) = 0$, $i = 1, 2$, and $\nabla f(0) = 0$.*

ii) *The ellipticity condition (5) shows that* $\nabla f(Z) \cdot Z$ *is at least of linear growth which implies*

$$a|Z| - b \le f(Z) \quad \text{for all } Z \in \mathbb{R}^n$$

and with some real numbers $a > 0$, b.

iii) *From the right-hand side of (5) we see that* f *is at most of quadratic growth and therefore (compare [Da], Lemma 2.2, p. 156)*

$$|\nabla f(Z)| \le c(1 + |Z|) \quad \text{for all } Z \in \mathbb{R}^n,$$

where c *is another positive constant.*

At this point we should give some examples to describe the class of energy densities which we have in mind by introducing Assumption 6.1.

Example 6.3. *With the notation of Assumption 6.1 we may take:*

i) *a linear growth integrand with* μ-*ellipticity,* $1 < \mu < 2$, *as given in Example 3.9: let*

$$\varphi(r) = \int_0^r \int_0^s \left(1 + t^2\right)^{-\frac{\mu}{2}} \, dt \, ds \,, \quad r \in \mathbb{R}_0^+ \,,$$

and choose $f_1(Z_1) = \varphi(|Z_1|)$.

ii) *The most elementary superlinear part is of power growth, i.e.*

$$f_2(Z_2) = \left(1 + |Z_2|^2\right)^{\frac{p}{2}}, \quad 1 < p \le 2 \,.$$

iii) *Anisotropic behavior of* f_2 *itself is not excluded: if* $Z_2 = (P, Q) \in \mathbb{R}^{n-k}$, $n - k \ge 2$, *then*

$$f_2(Z_2) = \left(1 + |P|^2\right)^{\frac{p}{2}} + \left(1 + |Q|^2\right)^{\frac{q}{2}}, \quad 1 < p < q \le 2$$

is admissible.

iv) *With the notation of iii) there is no need to assume the above decomposition of* f_2, *i.e. one may think of*

$$f_2(Z_2) = \left[1 + |P|^2 + \left(1 + |Q|^2\right)^{\frac{q}{p}}\right]^{\frac{p}{2}}.$$

We now come to a precise formulation of the main result of this chapter. Consider the Dirichlet problem

$$J[w] := \int_\Omega f(\nabla w) \, dx \to \min \quad \text{in } u_0 + \overset{\circ}{W}{}_1^1(\Omega) \,, \tag{$\mathcal{P}$}$$

where the boundary values u_0 are supposed to be of class $W_\infty^1(\Omega)$ (compare Remark 6.12 below for this choice). Since f_i, $i = 1, 2$, is at least of linear growth, we follow Chapter 2 and Chapter 4 to define generalized minimizers u^* of the problem $(\mathcal{P})$ by

$$u^* \in \mathcal{M} = \{u \in BV(\Omega) : u \text{ is the } L^1\text{-limit of a } J\text{-minimizing}$$

$$\text{sequence from } u_0 + \overset{\circ}{W}{}^1_1(\Omega)\} .$$

Remark 6.4. *We like to recall the alternatives discussed in Appendix A to define generalized minimizers for variational problems with linear growth.*

The main result of this section now is

Theorem 6.5. *Let f satisfy the general hypothesis, i.e. Assumption 6.1, and suppose that $u_0 \in W^1_\infty(\Omega)$. Then any generalized minimizer $u^* \in \mathcal{M}$ of the problem (P) is of class $C^{1,\alpha}(\Omega)$ for any $0 < \alpha < 1$. Moreover, the elements of $\mathcal{M}$ are uniquely determined up to a constant.*

We recall

$$Z = (Z_1, Z_2) \in \mathbb{R}^k \times \mathbb{R}^{n-k} , \quad f(Z) = f_1(Z_1) + f_2(Z_2)$$

from Assumption 6.1 and introduce the notation

$$w = w(x^1, x^2) , \quad x = (x^1, x^2) \in \Omega \subset \mathbb{R}^k \times \mathbb{R}^{n-k} ,$$

$$\nabla w = (\nabla_{x^1} w, \nabla_{x^2} w) \in \mathbb{R}^k \times \mathbb{R}^{n-k} ,$$

for any weakly differentiable function $w \colon \Omega \to \mathbb{R}$. Moreover, on account of the growth rate of f, the domain $u_0 + \overset{\circ}{W}{}^1_1(\Omega)$ of the energy J is replaced in the following by the anisotropic space

$$u_0 + \overset{\circ}{W}{}^1_{1,p}(\Omega) := \left\{ w \in u_0 + \overset{\circ}{W}{}^1_1(\Omega) : \nabla_{x^2} u \in L^p(\Omega; \mathbb{R}^{n-k}) \right\} .$$

Finally, recall the introductory remarks on duality theory given in Section 2.1.1, where now the representation formula for J reads as $(p' = p/(p-1))$:

$$J[w] = \sup_{\varkappa \in L^{\infty, p'}(\Omega; \mathbb{R}^n)} \left\{ \int_\Omega \varkappa \cdot \nabla w \, dx - \int_\Omega f_1^*(\varkappa_1) \, dx \right. \tag{7}$$

$$\left. - \int_\Omega f_2^*(\varkappa_2) \, dx \right\} , \quad w \in u_0 + \overset{\circ}{W}{}^1_{1,p}(\Omega) .$$

Here, for $1 \leq s, t \leq \infty$ the notation

$$\varkappa \in L^{s,t}(\Omega; \mathbb{R}^n) \Leftrightarrow \varkappa = (\varkappa_1, \varkappa_2) \in L^s(\Omega; \mathbb{R}^k) \times L^t(\Omega; \mathbb{R}^{n-k})$$

is introduced. Note that the assumptions of Proposition 2.1, [ET], p. 271, are clearly satisfied without a further specification of the upper growth rate of f_2. Thus, (7) is really follows as outlined in Section 2.1.1. The Lagrangian

$l(w, \varkappa)$ is defined for all $(w, \varkappa)$ in the class $(u_0 + \overset{\circ}{W}{}^1_{1,p}(\Omega)) \times L^{\infty,p'}(\Omega; \mathbb{R}^n)$ by the formula

$$l(w, \varkappa) := \int_\Omega \varkappa \cdot \nabla w \, dx - \sum_{i=1}^{2} \int_\Omega f_i^*(\varkappa_i) \, dx \,.$$

Remark 6.6. *If f_2 is of p growth, i.e.*

$$f_2(Z_2) = \left(1 + |Z_2|^2\right)^{\frac{p}{2}}, \quad 1 < p \leq 2, \tag{8}$$

for all $Z_2 \in \mathbb{R}^{n-k}$, then we may follow the lines of Chapter 2.1.1, define the dual functional $R: L^{\infty,p'}(\Omega; \mathbb{R}^n) \to \overline{\mathbb{R}}$,

$$R[\varkappa] := \inf_{w \in u_0 + \overset{\circ}{W}{}^1_{1,p}(\Omega)} l(w, \varkappa) \,,$$

and obtain the dual variational problem

$$\text{to maximize } R \text{ among all functions } \varkappa \in L^{\infty,p'}(\Omega; \mathbb{R}^n) \,. \tag{$\mathcal{P}^*$}$$

Again we have for any $\varkappa \in L^{\infty,p'}(\Omega; \mathbb{R}^n)$ (compare Remark 2.4)

$$R[\varkappa] = \begin{cases} -\infty & \text{if } \operatorname{div} \varkappa \neq 0 \,, \\ l(u_0, \varkappa) & \text{if } \operatorname{div} \varkappa = 0 \,. \end{cases}$$

With (8), the existence proof for a dual solution is standard (see [ET], Theorem 4.1, p 59), moreover we have

$$\inf_{w \in u_0 + \overset{\circ}{W}{}^1_{1,p}(\Omega)} J[w] = \sup_{\varkappa \in L^{\infty,p'}(\Omega;\mathbb{R}^n)} R[\varkappa] \,.$$

(In Chapter 2 these two results were obtained as a byproduct – compare Lemma 2.6 and Remark 2.9, i).)

We do not want to restrict to the consideration of energy densities satisfying (8) (see the examples discussed in 6.3). Then, due to the lack of continuity of J on $W^1_{1,p}(\Omega)$, the existence Theorem 4.1 of [ET], p. 59, is no longer applicable. Nevertheless, assuming

$$\operatorname{div} \varkappa = 0 \,, \quad \varkappa \in L^{\infty,p}(\Omega; \mathbb{R}^n) \,, \quad f_i^*(\varkappa^i) \in L^1(\Omega) \,, \quad i = 1, 2, \tag{9}$$

we again let

$$R[\varkappa] := l(u_0, \varkappa) \,.$$

Note that on account of $u_0 \in W^1_\infty(\Omega)$ and since we have (9), we do not have to suppose $\varkappa_2 \in L^{p'}(\Omega, \mathbb{R}^{n-k})$.

Given these preliminaries we are now going to give a *Proof of Theorem 6.5* which splits into four parts: regularization, integrability results, the study of a dual limit and the conclusion.

Step 1. (Regularization) Here the natural choice of the regularization is given by a Dirichlet perturbation, i.e. we let u_δ denote the unique solution of

$$J_\delta[w] := \int_\Omega f_\delta(\nabla w) \, dx \to \min \quad \text{in } u_0 + \overset{\circ}{W}{}^1_2(\Omega) \,, \qquad (\mathcal{P}_\delta)$$

where

$$f_\delta(Z) := f(Z) + \frac{\delta}{2} |Z|^2 \,, \quad \delta \in (0,1) \,.$$

Moreover, we introduce

$$\sigma_\delta := \nabla f_\delta(\nabla u_\delta) \,,$$

hence div $\sigma_\delta = 0$ follows by definition. As usual, u_δ satisfies

Lemma 6.7. *With the obvious changes in notation, the Caccioppoli-type inequalities as stated (for obstacle problems) in Lemma 5.20 are valid.*

Step 2. (Integrability results) The uniform higher integrability properties of the sequence $\{(u_\delta, \sigma_\delta)\}$ are summarized in

Lemma 6.8.

  i) *The sequence $\{\sigma_\delta\}$ is uniformly bounded in $L^{2,p}(\Omega; \mathbb{R}^n)$.*

 ii) *The sequence $\{\nabla u_\delta\}$ is uniformly bounded in $L^\infty_{loc}(\Omega; \mathbb{R}^n)$.*

iii) *The sequence $\{\sigma_\delta\}$ is uniformly bounded in $W^1_{2,loc}(\Omega; \mathbb{R}^n)$.*

iv) *There is a local constant c, independent of δ, such that*

$$\|\nabla^2 u_\delta\|_{L^2(\widehat\Omega; \mathbb{R}^{nn})} < c(\widehat\Omega)$$

  for any domain $\widehat\Omega \Subset \Omega$.

Proof. ad i). We have by definition

$$\sigma_\delta = (\sigma^1_\delta, \sigma^2_\delta) = \delta(\nabla_{x^1} u_\delta, \nabla_{x^2} u_\delta) + \left(\nabla f_1(\nabla_{x^1} u_\delta), \nabla f_2(\nabla_{x^2} u_\delta)\right) \,,$$

and on account of $J_\delta[u_\delta] \leq J_\delta[u_0] \leq J_1[u_0]$ we deduce the uniform bounds

$$\delta \int_\Omega |\nabla u_\delta|^2 \, dx \leq c \,, \quad \int_\Omega f_i(\nabla_{x^i} u_\delta) \, dx \leq c \,, \quad i = 1, 2 \,. \qquad (10)$$

Note that the first inequality of (10) implies (compare (7) of Section 2.1.2)

$$\|\delta \nabla u_\delta\|_{L^2(\Omega; \mathbb{R}^n)} \to 0 \quad \text{as } \delta \to 0 \,. \qquad (11)$$

Now the assertion for σ_δ^1 follows from (3), whereas (4) (together with (10)) gives a uniform bound for $\|\nabla_{x^2} u_\delta\|_{L^p(\Omega;\mathbb{R}^n)}$. As a consequence, we may apply Remark 6.2, *iii*), with the result

$$\int_\Omega |\nabla f_2(\nabla_{x^2} u_\delta)|^p \, \mathrm{d}x \le c \left\{ 1 + \int_\Omega |\nabla_{x^2} u_\delta|^p \, \mathrm{d}x \right\} \le c \, ,$$

which establishes the first claim (recall (11) and $p \le 2$).

ad *ii*). Local uniform gradient bounds for $\{u_\delta\}$ are proved in Theorem 5.22. In order to apply this theorem in the case of anisotropic linear/superlinear growth, we first observe that the Caccioppoli-type inequality Lemma 6.7 remains unchanged. Moreover, we have (q=) $2 < 4 - \mu$ by (6), hence relation (4), Section 5.1. Then Theorem 5.21 is valid whenever the choice $\alpha_0 = 0$ gives uniform integrability to start the iteration procedure. This however is an obvious consequence of $(2 - \mu)/2 < 1/2$ and the linear growth of f_1. Note that the analog of Theorem 5.21 then implies the corresponding version of Theorem 5.22, hence *ii*).

ad *iii*). The Cauchy-Schwarz inequality gives almost everywhere

$$|\nabla \sigma_\delta|^2 = D^2 f_\delta(\nabla u_\delta)(\partial_\gamma \nabla u_\delta, \partial_\gamma \sigma_\delta)$$

$$\le \left[D^2 f_\delta(\nabla u_\delta)(\partial_\gamma \nabla u_\delta, \partial_\gamma \nabla u_\delta) \right]^{\frac{1}{2}} \left[D^2 f_\delta(\nabla u_\delta)(\partial_\gamma \sigma_\delta, \partial_\gamma \sigma_\delta) \right]^{\frac{1}{2}} \, ,$$

and since $|D^2 f_\delta|$ is uniformly bounded there exists a constant, independent of δ, such that

$$|\nabla \sigma_\delta|^2 \le c D^2 f_\delta(\nabla u_\delta)(\partial_\gamma \nabla u_\delta, \partial_\gamma \nabla u_\delta) \, .$$

Hence, the claim follows from *ii*) and Lemma 6.7.

ad *iv*) Again we observe on account of *ii*) that for any domain $\widehat{\Omega} \Subset \Omega$

$$\int_{\widehat{\Omega}} (1 + |\nabla u_\delta|^2)^{-\frac{\mu}{2}} |\partial_\gamma \nabla u_\delta|^2 \, \mathrm{d}x \le \int_{\widehat{\Omega}} D^2 f_\delta(\nabla u_\delta)(\partial_\gamma \nabla u_\delta, \partial_\gamma \nabla u_\delta) \, \mathrm{d}x$$

$$\le c(\widehat{\Omega}) < \infty \, ,$$

and if we once more apply *ii*) on the left-hand side, then *iv*) is proved as well. □

Remark 6.9. *Note that we have a lack of integrability in i) which is due to the anisotropic behavior of the function f_2. If we restrict ourselves to the consideration of linear/p-growth energy densities as discussed in Remark 6.6, then ∇f_2 has a growth rate $p - 1$. Hence $\nabla f_2(\nabla_{x^2} u_\delta)$ is uniformly bounded in $L^{p'}(\Omega; \mathbb{R}^n)$, $p' = p/p - 1$. Moreover, f_2^* is of power p' and $f_2^*(\nabla f_2(\nabla_{x^2} u_\delta))$ is uniformly bounded in $L^1(\Omega)$. Besides the existence of a dual solution, this essentially enlarges the class of admissible boundary values (see Remark 6.12 below).*

Corollary 6.10.

i) *If u^* denotes a L^1-cluster point of the sequence $\{u_\delta\}$, then, after passing to a subsequence,*

$$\nabla u_\delta \to \nabla u^* \quad \text{almost everywhere in } \Omega \text{ as } \delta \to 0.$$

ii) *If σ denotes the corresponding $L^{2,p}$-cluster point of $\{\sigma_\delta\}$, then the limit version of the duality relation, i.e.*

$$\sigma = \nabla f(\nabla u^*)$$

holds almost everywhere.

iii) *In particular, σ is a mapping into the open set $U := \mathrm{Im}(\nabla f)$.*

Proof. Since $\{u_\delta\}$ is uniformly bounded in $L^\infty(\Omega)$ and on account of Lemma 6.8, *ii)*, we have as $\delta \to 0$

$$u_\delta \overset{*}{\rightharpoonup} u^* \quad \text{in } W^1_{\infty,loc}(\Omega) , \tag{12}$$

at least for a subsequence. Moreover, from Lemma 6.8, *ii)*, *iv)*, and Sobolev's embedding theorem we deduce the existence of a subsequence such that

$$\nabla u_\delta \overset{\delta \to 0}{\rightharpoonup} : w \quad \text{in } L^t_{loc}(\Omega; \mathbb{R}^n) , \quad t < \frac{2n}{n-2} .$$

With (12) we have $w = \nabla u^*$ and passing to another subsequence, if necessary, the first claim is proved. The second one immediately follows from $\sigma_\delta = \nabla f_\delta(\nabla u_\delta)$. It remains to show that U is an open set. This, however, is outlined in the proof of Lemma 2.14. $\square$

Step 3. (A dual limit) From now on a weak $L^{2,p}$-cluster point σ of the sequence $\{\sigma_\delta\}$ is fixed and (without relabeling) we always pass to subsequences if necessary. Let

$$\tau_\delta^i = \nabla f_i(\nabla_{x^i} u_\delta) , \quad i = 1, 2 ,$$

hence τ_δ is uniformly bounded in $L^{\infty,p}(\Omega; \mathbb{R}^n)$ and we may assume as $\delta \to 0$

$$\tau_\delta^1 \overset{*}{\rightharpoonup} \tau^1 \quad \text{in } L^\infty(\Omega; \mathbb{R}^k) , \quad \tau_\delta^2 \rightharpoonup \tau^2 \quad \text{in } L^p(\Omega; \mathbb{R}^{n-k}) .$$

Note that Corollary 6.10 immediately gives $\tau = \sigma$. The main properties of σ are summarized in the following lemma, where we should keep Remark 6.9 (compare Remark 6.6) in mind.

Lemma 6.11.

i) *The limit σ is R-admissible in the sense that $\mathrm{div}\,\sigma = 0$.*

ii) *The limit σ also satisfies the remaining assumptions of (9).*

iii) *We have $\inf\left\{ J[w] : w \in u_0 + \overset{\circ}{W}{}^1_{1,p}(\Omega) \right\} \leq R[\sigma]$.*

Proof. Passing to the limit $\delta \to 0$, i) follows from $\operatorname{div} \sigma_\delta = 0$.

ad ii). The duality relation for the conjugate function (compare (2), Section 2.1.1) reads as

$$\tau_\delta^i \cdot \nabla_{x^i} u_\delta - f_i^*(\tau_\delta^i) = f_i(\nabla_{x^i} u_\delta) , \quad i = 1, 2 .$$

At this point let us shortly discuss the global starting integrabilities: since u_δ is of class $W_2^1(\Omega)$ and since ∇f is at most of linear growth (see Remark 6.2, iii)), τ_δ is of class $L^2(\Omega; \mathbb{R}^n)$ and the same is true for σ_δ. Moreover, $f(\nabla u_\delta)$ is of class $L^1(\Omega)$ (even uniformly) and the duality relation proves L^1-integrability (of course not uniformly) for $f_i^*(\tau_\delta^i)$, $i = 1, 2$, as well. As a consequence, the expressions of inequality (13) below are well defined. We write (using the definition of σ_δ together with $\operatorname{div} \sigma_\delta = 0$)

$$J[u_\delta] = \sum_{i=1}^{2} \int_\Omega \left[\tau_\delta^i \cdot \nabla_{x^i} u_\delta - f_i^*(\tau_\delta^i) \right] \, dx$$

$$= -\delta \int_\Omega |\nabla u_\delta|^2 \, dx + \sum_{i=1}^{2} \int_\Omega \left[\sigma_\delta^i \cdot \nabla_{x^i} u_\delta - f_i^*(\tau_\delta^i) \right] \, dx \qquad (13)$$

$$= -\delta \int_\Omega |\nabla u_\delta|^2 \, dx + \sum_{i=1}^{2} \int_\Omega \left[\sigma_\delta^i \cdot \nabla_{x^i} u_0 - f_i^*(\tau_\delta^i) \right] \, dx .$$

Let us have a closer look at (13): obviously $J[u_\delta] + \delta \int_\Omega |\nabla u_\delta|^2 \, dx$ is uniformly bounded and

$$\left| \int_\Omega \sigma_\delta \cdot \nabla u_0 \, dx \right| \leq c$$

(independent of δ) follows from Lemma 6.8, i), and the smoothness assumptions on u_0. Moreover, by the duality relation and by (3), (10)

$$\int_\Omega f_1^*(\tau_\delta^1) \, dx = \int_\Omega \left(\tau_\delta^1 \cdot \nabla_{x^1} u_\delta - f_1(\nabla_{x^1} u_\delta) \right) \, dx \leq c , \qquad (14)$$

again with a constant which does not depend on δ. Thus, the representation formula (13) gives a positive number c such that for any $\delta > 0$

$$\int_\Omega f_2^*(\tau_\delta^2) \, dx \leq c . \qquad (15)$$

Note that the integrability established in Lemma 6.8, i), is not sufficient to imply (15). Once (14) and (15) are established, the second claim follows from Fatou's lemma and Corollary 6.10, i).

ad iii). Now that the limit σ is seen to be R-admissible, the third claim follows from (13) exactly as in the proof of Lemma 2.6

$$\inf \left\{ J[w] : w \in u_0 + \overset{\circ}{W}{}_{1,p}^1(\Omega) \right\} \leq J[u_\delta]$$

$$= -\delta \int_\Omega |\nabla u_\delta|^2 \, dx + \sum_{i=1}^{2} \int_\Omega \left[\sigma_\delta^i \cdot \nabla_{x^i} u_0 - f_i^*(\tau_\delta^i) \right] \, dx .$$

Passing to the limit $\delta \to 0$ we obviously have (recall Lemma 6.8, i))

$$\sum_{i=1}^{2} \int_{\Omega} \sigma_{\delta}^{i} \cdot \nabla_{x^i} u_0 \, \mathrm{d}x \to \sum_{i=1}^{2} \int_{\Omega} \sigma^{i} \cdot \nabla_{x^i} u_0 \, \mathrm{d}x \, ,$$

moreover,

$$\limsup_{\delta \to 0} - \sum_{i=1}^{2} \int_{\Omega} f_i^*(\tau_{\delta}^i) \, \mathrm{d}x \leq - \sum_{i=1}^{2} \liminf_{\delta \to 0} \int_{\Omega} f_i^*(\tau_{\delta}^i) \, \mathrm{d}x \, .$$

As in ii), Fatou's lemma proves the last assertion. $\quad \square$

Remark 6.12. *Let us shortly discuss the choice of u_0. The assumption $u_0 \in L^{\infty}(\Omega)$ is needed to obtain uniformly bounded solutions u_δ, hence we may apply the results of Section 5.2 and obtain uniform local a priori gradient bounds.*

If the boundary values u_0 are not supposed to be in addition of class $W^1_{\infty}(\Omega)$, then we may pass to an approximating sequence u_0^m and regularize with respect to these boundary values. With the obvious changes in notation, it is to verify in this case that $(i = 1, 2)$

$$\int_{\Omega} \sigma_{\delta(m)}^{i} \cdot \nabla_{x^i} u_0^m \, \mathrm{d}x \to \int_{\Omega} \sigma^{i} \cdot \nabla_{x^i} u_0 \, \mathrm{d}x \quad \text{as } m \to \infty \, ,$$

where $\delta(m)$ is chosen sufficiently small. By assumption (3) we know that $|\nabla f_1| \leq A$, and it is sufficient to suppose $\nabla_{x^1} u_0 \in L^1(\Omega; \mathbb{R}^k)$ in order to prove the above convergence for $i = 1$. In the case $i = 2$ we recall Remark 6.9, hence we at least have to assume that $\nabla_{x^2} u_0 \in L^{p'}(\Omega; \mathbb{R}^{n-k})$.

Up to now it is not proved that $\{u_\delta\}$ is a J-minimizing sequence of class $u_0 + \overset{\circ}{W}{}^1_{1,p}(\Omega)$. Anyhow, we have the Euler equation of Lemma 6.11, i), hence

Corollary 6.13. *The limit σ fixed as above is of class $C^{0,\alpha}(\Omega; \mathbb{R}^n)$ for any $0 < \alpha < 1$.*

Proof. Corollary 6.10 yields

$$\int_{\Omega} \nabla f(\nabla u^*) \cdot \nabla \varphi \, \mathrm{d}x = 0 \quad \text{for all } \varphi \in C_0^1(\Omega)$$

and the standard arguments sketched at the end of Section 3.3.4 complete the proof. $\quad \square$

Step 4. (Conclusion) Roughly speaking it remains to derive an appropriate minimax inequality as given in Appendix A.3 although we do not know whether σ is a solution of the dual problem.

Fix $u \in \mathcal{M}$, i.e. consider a J-minimizing sequence $\{u_m\}$ in $u_0 + \overset{\circ}{W}{}^1_{1,p}(\Omega)$ such that as $m \to \infty$

$$u_m \overset{L^{n/(n-1)}}{\rightharpoonup} u , \quad u_m \overset{L^1}{\to} u .$$

For the sake of notational simplicity assume now that $\Omega = B = B_1(0)$, the general case is handled by an additional covering argument. Moreover, we still consider σ as discussed in Step 3 and let $\sigma_\rho := \sigma(\rho x)$ for $0 < \rho < 1$. Finally, for $\lambda \in C_0^\infty(\Omega; \mathbb{R}^n)$ and a real number t, $|t|$ sufficiently small, we let $\chi_\rho = \sigma_\rho + t\lambda(\rho x)$. Then χ_ρ is admissible to obtain (recall (7))

$$J[u_m] = \sup_{\varkappa \in L^{\infty,p'}(\Omega;\mathbb{R}^n)} l(u_m, \varkappa) \geq l(u_m, \chi_\rho) \tag{16}$$

and we may write (div $\sigma_\rho = 0$)

$$l(u_m, \chi_\rho) = t \int_\Omega \operatorname{div} \lambda(\rho x)(u_0 - u_m)\,\mathrm{d}x$$
$$- \sum_{i=1}^2 \int_\Omega f_i^*(\chi_\rho^i)\,\mathrm{d}x + \int_\Omega \chi_\rho \cdot \nabla u_0 \,\mathrm{d}x . \tag{17}$$

Passing to the limit $\rho \to 1$ we immediately obtain the convergence of the first integral on the right-hand side of (17). Letting $\chi = \sigma + t\lambda$, the last integral is seen to converge with limit $\int_\Omega \chi \cdot \nabla u_0 \,\mathrm{d}x$. The convergence of the remaining integrals follows from $f_i^*(\chi^i) \in L^1(\Omega)$, $i = 1, 2$, (which is proved in Lemma 6.11, ii)) together with a standard reasoning (compare [Alt], Lemma 1.16, p. 18).

Next we combine (16), (17) and pass to the limit $m \to \infty$ with the result

$$\inf_{w \in u_0 + \overset{\circ}{W}{}^1_{1,p}(\Omega)} J[w]$$
$$\geq t \int_\Omega \operatorname{div} \lambda (u_0 - u)\,\mathrm{d}x - \sum_{i=1}^2 \int_\Omega f_i^*(\chi^i)\,\mathrm{d}x + \int_\Omega \chi \cdot \nabla u_0 \,\mathrm{d}x .$$

If we additionally observe that (recall Lemma 6.11, iii))

$$\inf_{w \in u_0 + \overset{\circ}{W}{}^1_{1,p}(\Omega)} J[w] \leq R[\sigma]$$
$$= \int_\Omega \sigma \cdot \nabla u_0 \,\mathrm{d}x - \sum_{i=1}^2 \int_\Omega f_i^*(\sigma^i)\,\mathrm{d}x$$

then we obtain the variational inequality

$$t \int_\Omega \operatorname{div} \lambda (u_0 - u)\,\mathrm{d}x - \sum_{i=1}^2 \int_\Omega f_i^*(\chi^i)\,\mathrm{d}x + \int_\Omega \chi \cdot \nabla u_0 \,\mathrm{d}x \leq R[\sigma] .$$

Now we proceed exactly as in the proof of Theorem A.9, i.e. we recall the definition of χ and we arrive at

$$- \int_{\text{spt}\,\lambda} t\,(\text{div}\,\lambda)\, u \,\text{d}x \leq \sum_{i=1}^{2} \int_{\text{spt}\,\lambda} \left(f_i^*(\sigma^i + t\lambda^i) - f_i^*(\sigma^i) \right) \text{d}x \,.$$

Hence, dividing through $t > 0$ and passing to the limit $t \to 0$ we get

$$- \int_{\text{spt}\,\lambda} (\text{div}\,\lambda)\, u \,\text{d}x \leq \sum_{i=1}^{2} \int_{\text{spt}\,\lambda} \nabla f_i^*(\sigma^i) \cdot \lambda^i \,\text{d}x \,,$$

i.e., by definition, the first weak derivative of u is given by $(\nabla f_1^*(\sigma^1), \nabla f_2^*(\sigma^2))$. Since σ is of class $C^{0,\alpha}$ (see Corollary 6.13) and since σ is a mapping into the open set $\text{Im}(\nabla f)$ (this was proved in Corollary 6.10, *iii*)) we obtain Hölder continuity of the derivatives of u. Moreover, each $v \in \mathcal{M}$ satisfies $\nabla v = \nabla f^*(\sigma)$ and the uniqueness (up to a constant) of generalized minimizers is proved as well. $\square$

Remark 6.14. *Note that in Lemma 6.11 it is actually proved that (after passing to another subsequence)*

$$\lim_{\delta \to 0} J[u_\delta] \leq R[\sigma] \,.$$

Moreover, choosing $\lambda \equiv 0$ in Step 4, we see that

$$R[\sigma] \leq \inf_{w \in u_0 + \overset{\circ}{W}{}^1_{1,p}(\Omega)} J[w] \leq \lim_{\delta \to 0} J[u_\delta] \leq R[\sigma] \,,$$

thus $\{u_\delta\}$ in fact provides a J-minimizing sequence.

Remark 6.15. *We like to finish with the remark that the results of this chapter are somewhat surprising in the following sense: since the function f_1 is merely of linear growth, we expect difficulties while studying the deformation gradient ∇u, whereas the stress tensor usually admits a clear interpretation as the unique solution of the dual problem. In contrast to this expectation, generalized minimizers are uniquely determined up to a constant, and it is not clear whether the dual problem does even admit a solution (which is due to the possible anisotropy of f_2). Nevertheless, we have found a dual limit which satisfies the stress-strain relation and such that $R[\sigma]$ realizes $\inf J$, i.e. in some sense we have found a "local stress tensor". If global higher integrability of σ holds true, then this terminology coincides with the usual notion of a dual solution.*

A

Some remarks on relaxation

Although we do not want to contribute to the theory of relaxation, several results from this field serve as important tools for our purposes. For instance, the relaxed functional $\hat{J}$ was introduced in Section 2.3.1 in order to refer to the regularity results of Anzellotti and Giaquinta. The corresponding relaxed minimization problem $(\hat{\mathcal{P}})$ – involving the Dirichlet boundary data u_0 via the space $BV_{u_0}(\Omega; \mathbb{R}^N)$ (see (1) for the definition) – is discussed in [GMS1] for integrands with minimal surface structure.

We follow these lines in Section A.1 and show that any generalized minimizer $u^* \in \mathcal{M}$ is a solution of $(\hat{\mathcal{P}})$ and vice versa. Here we have to recall the notion of the set $\mathcal{M}$ as given, for instance, in the introduction. Of course the above claim once more proves Theorem 2.17 (compare Remark 2.16).

A (formally) different approach to the notion of a relaxed minimization problem is usually preferred in the theory of perfect plasticity following the lines of Seregin ([Se1], see [FuS2] for a complete list of references) and Strang and Temam ([ST]). In Section A.2 the basic definition is introduced. For experts in the theory of relaxation it might be obvious that both points of view lead to the same result but we did not find a rigorous proof in the literature. So we decided in [BF4] to sketch the main arguments, where we rely on the results of [DT] (see also [Te]).

Finally, in Section A.3 we prove two uniqueness results for generalized minimizers. The first one is taken from [BF3] with a main idea which is already given in [Se6].

Throughout this appendix Ω and $\hat{\Omega}$, $\Omega \Subset \hat{\Omega}$, denote bounded Lipschitz domains. Given boundary values $u_0 \in W_1^1(\Omega; \mathbb{R}^N)$ we extend u_0 to the domain $\hat{\Omega}$ such that $u_0 \in \overset{\circ}{W}{}_1^1(\hat{\Omega}; \mathbb{R}^N)$ and let

$$BV_{u_0}(\Omega; \mathbb{R}^N) := \left\{ u \in BV(\hat{\Omega}; \mathbb{R}^N) : u = u_0 \text{ on } \hat{\Omega} - \Omega \right\}. \tag{1}$$

Concerning the integrand f we do not need the full strength of Assumption 2.1 – it is sufficient to suppose

Assumption A.1. *The integrand* $f\colon \mathbb{R}^{nN} \to [0, \infty)$ *is strictly convex (in the sense of definition), of linear growth, i.e.*

$$a|Z| - b \leq f(Z) \leq A|Z| + B \quad \text{for all } Z \in \mathbb{R}^{nN}$$

with some positive constants a, A, b, B, *and satisfies* $f(0) = 0$.

A.1 The approach known from the minimal surface case

We start by recalling the definition of the relaxed functional $\hat{J}$ (see Definition 2.19 and Proposition 2.21)

$$\hat{J}[w, \hat{\Omega}] = \int_{\hat{\Omega}} f(\nabla^a w) \, dx + \int_{\hat{\Omega}} f_\infty \left(\frac{\nabla^s w}{|\nabla^s w|} \right) d|\nabla^s w| \,, \quad w \in BV(\Omega; \mathbb{R}^N) \,.$$

Given this representation, the lower semicontinuity theorem of Reshetnyak (see [Re]) immediately implies the existence result (compare [GMS1], Theorem 1.3)

Theorem A.2. *There exists a minimum point for the problem*

$$\hat{J}[w, \hat{\Omega}] \to \min \quad in \quad BV_{u_0}(\Omega; \mathbb{R}^N) \,.$$

Now we refer to [AFP], Theorem 3.77, p. 171, stating the well known relation

$$\nabla^s w \llcorner \partial\Omega = (u_0 - w) \otimes \nu \, \mathcal{H}^{n-1} \llcorner \partial\Omega \quad \text{for all } w \in BV_{u_0}(\Omega; \mathbb{R}^N) \,,$$

where ν denotes the unit outward normal to $\partial\Omega$. We obtain for any $w \in BV_{u_0}(\Omega; \mathbb{R}^N)$

$$
\begin{aligned}
\hat{J}[w, \hat{\Omega}] = &\int_\Omega f(\nabla^a w) \, dx + \int_\Omega f_\infty \left(\frac{\nabla^s w}{|\nabla^s w|} \right) d|\nabla^s w| \\
&+ \int_{\partial\Omega} f_\infty((u_0 - w) \otimes \nu) \, d\mathcal{H}^{n-1} + \int_{\hat{\Omega}-\Omega} f(\nabla u_0) \, dx \,.
\end{aligned}
\tag{2}
$$

The last integral on the right-hand side of (2) is constant, thus we may drop this term and introduce the energy

$$
\begin{aligned}
K[w] := &\int_\Omega f(\nabla^a w) \, dx + \int_\Omega f_\infty \left(\frac{\nabla^s w}{|\nabla^s w|} \right) d|\nabla^s w| \\
&+ \int_{\partial\Omega} f_\infty((u_0 - w) \otimes \nu) \, d\mathcal{H}^{n-1} \,, \quad w \in BV(\Omega; \mathbb{R}^N) \,.
\end{aligned}
\tag{3}
$$

This implies the representation

$$\hat{J}[\hat{w}, \hat{\Omega}] = K[w_{|\Omega}] + const \quad \text{for any } \hat{w} \in BV_{u_0}(\Omega; \mathbb{R}^N) \,. \tag{4}$$

Conversely, given $w \in BV(\Omega; \mathbb{R}^N)$, the extension $\hat{w}$ via u_0 to the domain $\hat{\Omega}$,

$$\hat{w} := \begin{cases} u_0 \text{ on } \hat{\Omega} - \Omega \,, \\[2mm] w \text{ on } \Omega \,, \end{cases} \tag{5}$$

is of class $BV_{u_0}(\Omega; \mathbb{R}^N)$ satisfying

$$\hat{J}[\hat{w}; \hat{\Omega}] = K[w] + const \,, \tag{6}$$

where the constant is the same as in (4). Thus, with (2)–(6) there is a one-to-one correspondence between $\hat{J}$ and K.

The main properties of the functional K defined in (3) are summarized in

Theorem A.3.

i) *The minimization problem*

$$K[w] \to \min \quad in \ BV(\Omega; \mathbb{R}^N) \tag{$\hat{\mathcal{P}}$}$$

admits at least one solution.

ii) $\displaystyle \inf_{u_0 + \overset{\circ}{W}{}^1_1(\Omega;\mathbb{R}^N)} J[w] = \inf_{BV(\Omega;\mathbb{R}^N)} K[w] \,.$

iii) $u^* \in \mathcal{M} \Leftrightarrow u^*$ *is K-minimizing.*

Proof. *i)* is immediate by Theorem A.2 and the correspondence (2)–(6).

Assertion *ii)* is outlined in the minimal surface case in [Giu2], Proposition 14.3, p. 161. In the same spirit we first observe that for $w \in u_0 + \overset{\circ}{W}{}^1_1(\Omega; \mathbb{R}^N)$ obviously $K[w] = J[w]$ holds, hence it is to prove that the left-hand side of *ii)* is less than or equal to the right-hand side. The validity of this inequality can be established as a corollary of the continuity Proposition 2.22 together with the density Lemma B.2. Although the density lemma looks quite familiar, a detailed proof can only hardly be found and we prefer to give an explicit construction in Appendix B.1.

ad *iii)*: exactly as in part *ii)* we see that $u^* \in \mathcal{M}$ whenever u^* is K-minimizing. Conversely, consider $u^* \in \mathcal{M}$ as the L^1-limit of some J-minimizing sequence $\{u_m\}$ from $u_0 + \overset{\circ}{W}{}_1^1(\Omega; \mathbb{R}^N)$. Since K is lower semi-continuous with respect to this kind of convergence (see Proposition 2.20, *i)*), we find

$$K[u^*] \leq \liminf_{m \to \infty} K[u_m] = \liminf_{m \to \infty} J[u_m] = \inf_{w \in u_0 + \overset{\circ}{W}{}^1_1(\Omega;\mathbb{R}^N)} J[w] \,,$$

and by *ii)* u^* is seen to be K-minimizing. $\square$

A.2 The approach known from the theory of perfect plasticity

In order to introduce the notion of a relaxed problem via some suitable Lagrangian function, we recall the definition of $l(w, \varkappa)$ given in Section 2.1.1, which can be written as

$$\tilde{l}(w, \varkappa) = \int_{\Omega} \operatorname{div} \varkappa (u_0 - w) \, \mathrm{d}x - \int_{\Omega} f^*(\varkappa) \, \mathrm{d}x + \int_{\Omega} \varkappa : \nabla u_0 \, \mathrm{d}x .$$

Here we use the symbol $\tilde{l}(w, \varkappa)$ instead of $l(w, \varkappa)$ to indicate that we now take $w \in BV(\Omega; \mathbb{R}^N)$ and

$$\varkappa \in \mathcal{U} := \left\{ \tau \in L^\infty(\Omega; \mathbb{R}^{nN}) : \operatorname{div} \tau \in L^n(\Omega; \mathbb{R}^N) \right\} .$$

Motivated by the representation formula (3), Section 2.1.1, we are led to the

Definition A.4. *On the space $BV(\Omega; \mathbb{R}^N)$, the functional $\tilde{J}$ is given by*

$$\tilde{J}[w] := \sup_{\varkappa \in \mathcal{U}} \tilde{l}(w, \varkappa) .$$

The next result is essentially due to Seregin (see, e.g. [Se1], we just added the fact that $\tilde{J}$-minimizers lie in the set $\mathcal{M}$), but with the help of Theorem A.6 it can also be reduced to Theorem A.3.

Theorem A.5.

i) The minimization problem

$$\tilde{J}[w] \to \min \quad in \ BV(\Omega; \mathbb{R}^N) \tag{$\tilde{\mathcal{P}}$}$$

admits at least one solution.

ii) $\displaystyle \inf_{u_0 + \overset{\circ}{W}{}^1_1(\Omega; \mathbb{R}^N)} J[w] = \inf_{BV(\Omega; \mathbb{R}^N)} \tilde{J}[w] .$

iii) $u^* \in \mathcal{M} \Leftrightarrow u^*$ *is $\tilde{J}$-minimizing.*

So, according to Theorem A.3 and Theorem A.5, the problem $(\hat{\mathcal{P}})$ as well as the problem $(\tilde{\mathcal{P}})$ is a suitable relaxed version of the original problem $(\mathcal{P})$ in the sense that we have the properties stated in *ii)* and *iii)*. Now, in order to get a complete picture of the situation, we formulate our main result.

Theorem A.6. *Assume that f satisfies the hypotheses stated in Assumption A.1. Then the functionals K and $\tilde{J}$ defined above coincide on the space $BV(\Omega; \mathbb{R}^N)$.*

Before going into the proof of the identification theorem we have to show the following auxiliary result

Lemma A.7. *Under the Assumption A.1 on f the conjugate function is essentially smooth, i.e. f* is a proper convex function and for $D = \text{int}(\text{dom} f^*)$ we have*

i) D is not empty;

ii) f* is differentiable throughout D;

iii) $\lim_{i \to \infty} |\nabla f^*(Q_i)| = +\infty$ *whenever $\{Q_i\}$ is a sequence in D converging to a boundary point Q of D.*

Proof of Lemma A.7. As a convex and finite function on $\mathbb{R}^{nN}$, f is continuous (see [Ro], Corollar 10.1.1, p. 83). For a proper convex function closedness is the same as lower semicontinuity ([Ro], p. 52), hence in particular f is closed. As a strictly convex function, f clearly is essentially strictly convex, thus we may apply Theorem 26.3, p. 253, of [Ro] to see that f^* is essentially smooth.
□

Remark A.8. *From the linear growth condition on f it follows that dom f* is a bounded set, compare, e.g. [DT], Section 1.2, moreover $f^* \geq 0$ and $f^*(0) = 0$ on account of the corresponding conditions on f.*

We now come to the *Proof of Theorem A.6.* Let

$$L(c) := \left\{ Q \in \mathbb{R}^{nN} : f^*(Q) \leq c \right\}, \quad c \in \mathbb{R}.$$

We claim that the Definition A.4 of $\tilde{J}$ can be rewritten in the following form: for any $w \in BV(\Omega; \mathbb{R}^N)$ the identity

$$\tilde{J}[w] = \sup \left\{ \tilde{l}(w, \varkappa) : \varkappa \in C^\infty(\overline{\Omega}; \mathbb{R}^{nN}), \right.$$
$$\left. \varkappa \in L(c) \text{ on } \overline{\Omega} \text{ for some } c = c(\varkappa) \in \mathbb{R} \right\} \tag{7}$$

is valid. In fact, the first part of Lemma A.7 shows that $Q_0 \in D$ may be fixed in the following. Moreover, given $w \in BV(\Omega; \mathbb{R}^N)$ and $\varepsilon > 0$, a tensor $\varkappa = \varkappa_\varepsilon \in \mathcal{U}$ is chosen such that

$$\tilde{J}[w] - \tilde{l}(w, \varkappa) < \varepsilon,$$

in particular we may assume that $\varkappa(x) \in \text{dom} f^*$ almost everywhere. For a sequence $\{\lambda_k\}$ in $(0, 1)$, $\lambda_k \to 1$ as $k \to \infty$, we then let

$$\varkappa_k := (1 - \lambda_k) Q_0 + \lambda_k \varkappa.$$

Since Q_0 belongs to the interior of $\text{dom} f^*$, we can find an open ball B around Q_0 completely contained in D. Let $C = C(x)$ denote the union of all segments $\overline{P\varkappa(x)}$ with $P \in B$. Clearly C is contained in the convex set $\text{dom} f^*$, and any point $Q \in \overline{Q_0 \varkappa(x)}$ – different from $\varkappa(x)$ in the case $\varkappa(x) \in \partial\mathrm{dom}\, f^*$ – belongs to the interior of $\mathrm{dom}\, f^*$, thus $\varkappa_k(x) \in D$ for almost all x and any k. Next, we prove the existence of real numbers $\gamma_k > 0$ such that

$$\mathrm{dist}\,(\varkappa_k, \partial\mathrm{dom}\, f^*) \geq \gamma_k \quad \text{almost everywhere and for any } k \in \mathbb{N}. \tag{8}$$

In fact, with the notation

$$M_\lambda := \left\{ (1 - \lambda)\, Q_0 + \lambda Q : Q \in \mathrm{dom}\, f^* \right\}, \quad 0 < \lambda < 1,$$

assume by contradiction that there exists $0 < \lambda < 1$ and a sequence $\{P_m\}$ in M_λ such that $P_m \to \bar{P} \in \partial\mathrm{dom}\, f^*$ as $m \to \infty$. By definition

$$P_m = (1 - \lambda)\, Q_0 + \lambda Q_m$$

for some sequence $\{Q_m\}$ in $\mathrm{dom}\, f^*$ and we may assume that $Q_m \to \bar{Q} \in \overline{\mathrm{dom}\, f^*}$ as $m \to \infty$. This immediately yields

$$\partial\mathrm{dom}\, f^* \ni \bar{P} = (1 - \lambda)\, Q_0 + \lambda\bar{Q}\,.$$

On the other hand, Theorem 6.1, p. 45 of [Ro], implies with the above representation that $\bar{P} \in \mathrm{int}\,(\overline{\mathrm{dom}\, f^*}) = \mathrm{int}\,(\mathrm{dom}\, f^*)$, hence by contradiction (8) is proved. Remarking that

$$K_k := \left\{ Q \in \mathrm{dom}\, f^* : \mathrm{dist}\,(Q, \partial\mathrm{dom}\, f^*) \geq \gamma_k \right\}$$

is a compact set contained in $\mathrm{int}\,(\mathrm{dom}\, f^*)$, continuity of f^* on $\mathrm{int}\,(\mathrm{dom}\, f^*)$ implies the existence of a real number c_k such that $f^* \leq c_k$ on K_k, and recalling (8) we deduce

$$\varkappa_k \in L(c_k) \quad \text{almost everywhere.} \tag{9}$$

From

$$f^*(\varkappa_k) \leq (1 - \lambda_k)\, f^*(Q_0) + \lambda_k\, f^*(\varkappa)$$

and the choice of $\varkappa$ we get

$$\tilde{l}(w, \varkappa_k) \geq \lambda_k \tilde{l}(w, \varkappa) + (1 - \lambda_k) \left(\int_\Omega Q_0 : \nabla u_0 \, dx - |\Omega|\, f^*(Q_0) \right)$$

$$\geq \lambda_k \left(\tilde{J}[w] - \varepsilon \right) + (1 - \lambda_k) \left(\int_\Omega Q_0 : \nabla u_0 \, dx - |\Omega|\, f^*(Q_0) \right),$$

therefore

$$\tilde{J}[w] \leq \tilde{l}(w, \varkappa_k) + 2\varepsilon \tag{10}$$

for all $k \gg 1$. Let us fix such an integer k. In order to verify our claim (7), we apply a modification of the approximation Lemma A.1.1 of [FuS2] to the tensor $\varkappa_k$. The modification is necessary since it is not clear that the construction provided in [FuS2] preserves condition (9). Anyhow, with Lemma B.3 we have a rigorous proof that there exists a sequence $\{\varkappa_{k,m}\}_{m\in\mathbb{N}}$ in $C^\infty(\overline{\Omega};\mathbb{R}^{nN})$ with values in $L(c_k)$ and such that

$$\tilde{l}(w,\varkappa_{k,m}) \overset{m\to\infty}{\Rightarrow} \tilde{l}(w,\varkappa_k) \, ,$$

where $\int_\Omega f^*(\varkappa_{k,m})\,\mathrm{d}x \to \int_\Omega f^*(\varkappa_k)\,\mathrm{d}x$ as $m\to\infty$ follows from the almost everywhere convergence of $\varkappa_{k,m}$ together with the level-set property. Hence we deduce from (10) the inequality

$$\tilde{J}[w] \leq \tilde{l}(w,\varkappa_{k,m}) + 3\varepsilon \, ,$$

at least for $m \gg 1$, and (7) is established.

Next, observe that the supremum in (7) may be taken over all $\varkappa = \hat{\varkappa}_{|\overline{\Omega}}$, $\hat{\varkappa} \in C_0^\infty(\hat{\Omega};\mathbb{R}^{nN})$, where the level-set property remains unchanged (see Remark B.4). Moreover, $w \in BV(\Omega;\mathbb{R}^N)$ again is extended to $\hat{w} \in BV_{u_0}(\Omega;\mathbb{R}^N)$ (see (5), Appendix A.1) and we get the representation

$$\tilde{J}[w] = \sup\left\{ \int_{\hat{\Omega}} \operatorname{div}\varkappa(u_0-\hat{w})\,\mathrm{d}x - \int_\Omega f^*(\varkappa)\,\mathrm{d}x + \int_\Omega \varkappa:\nabla u_0\,\mathrm{d}x : \right.$$
$$\left. \varkappa \in C_0^\infty(\hat{\Omega};\mathbb{R}^{nN}), \quad \varkappa \in L(c) \text{ on } \hat{\Omega} \text{ for some } c = c(\varkappa) \in \mathbb{R} \right\} \, .$$

An integration by parts leads to

$$\tilde{J}[w] = \sup\left\{ \int_{\overline{\Omega}} \varkappa:\nabla\hat{w} - \int_\Omega f^*(\varkappa)\,\mathrm{d}x : \quad \varkappa \in C_0^\infty(\hat{\Omega};\mathbb{R}^{nN}), \right.$$
$$\left. \varkappa \in L(c) \text{ on } \hat{\Omega} \text{ for some } c = c(\varkappa) \in \mathbb{R} \right\} \tag{11}$$

being valid for any $w \in BV(\Omega;\mathbb{R}^N)$. Clearly, a smoothing argument shows, that in (11) the space $C_0^\infty(\hat{\Omega};\mathbb{R}^{nN})$ can be replaced by $C_0^0(\hat{\Omega};\mathbb{R}^{nN})$.

Let us fix $w \in BV(\Omega;\mathbb{R}^N)$ and a tensor $\varkappa \in C_0^0(\hat{\Omega};\mathbb{R}^{nN})$. If $\varkappa \in L(c)$, then $f^* \circ \varkappa$ is of class $L^1(\hat{\Omega})$. Conversely, let us assume that $f^* \circ \varkappa$ is integrable on $\hat{\Omega}$ implying $\varkappa(x) \in \operatorname{dom} f^*$. Then, using the arguments presented after the statement (7), we can construct tensors $\varkappa_k := (1-\lambda_k)Q_0 + \lambda_k\varkappa$ as before which we multiply by some function $\eta \in C_0^0(\hat{\Omega})$, $0 \leq \eta \leq 1$, $\eta \equiv 1$ on a neighborhood of $\overline{\Omega}$. The tensors $\chi_k := \eta\varkappa_k$ are of class $C_0^0(\hat{\Omega};\mathbb{R}^{nN})$ and

$$f^*(\chi_k) = f^*((1-\eta)0 + \eta\varkappa_k) \leq \eta f^*(\varkappa_k) \leq c_k$$

provided that $\varkappa_k \in L(c_k)$. We have

$$\int_{\overline{\Omega}} \chi_k:\nabla\hat{w} = \int_{\overline{\Omega}} \varkappa_k:\nabla\hat{w} \overset{k\to\infty}{\Rightarrow} \int_{\overline{\Omega}} \varkappa:\nabla\hat{w} \, ,$$

on Ω we estimate

$$0 \le f^*(\chi_k) = f^*(\varkappa_k) \le (1 - \lambda_k) f^*(Q_0) + \lambda_k f^*(\varkappa)$$

$$\le \max \{ f^*(Q_0), f^*(\varkappa) \} \in L^1(\Omega) \,,$$

hence, passing to the limit $k \to \infty$, $\int_\Omega f^*(\chi_k) \, dx \to \int_\Omega f^*(\varkappa) \, dx$ follows from dominated convergence and $f^*(\chi_k) \to f^*(\varkappa)$ almost everywhere on Ω. Altogether we find

$$\int_{\overline{\Omega}} \varkappa : \nabla \hat{w} - \int_\Omega f^*(\varkappa) \, dx = \lim_{k \to \infty} \left\{ \int_{\overline{\Omega}} \chi_k : \nabla \hat{w} - \int_\Omega f^*(\chi_k) \, dx \right\},$$

and (11) implies

$$\tilde{J}[w] = \sup \left\{ \int_{\hat{\Omega}} 1_{\overline{\Omega}} \varkappa : \nabla \hat{w} - \int_\Omega f^*(\varkappa) \, dx : \ \varkappa \in C_0^0 (\hat{\Omega}; \mathbb{R}^{nN}), \right.$$

$$\left. f^* \circ \varkappa \in L^1(\hat{\Omega}) \right\}$$

$$\ge \sup \left\{ \int_{\hat{\Omega}} 1_{\overline{\Omega}} \varkappa : \nabla \hat{w} - \int_{\hat{\Omega}} f^*(\varkappa) \, dx : \ \varkappa \in C_0^0 (\hat{\Omega}; \mathbb{R}^{nN}), \right.$$

$$\left. f^* \circ \varkappa \in L^1(\hat{\Omega}) \right\}, \tag{12}$$

the inequality being a consequence of $f^* \ge 0$. Consider a tensor $\varkappa$ which realizes the first supremum up to a given $\varepsilon > 0$. Then we let $\eta_k \in C_0^0(\hat{\Omega})$, $0 \le \eta_k \le 1$, such that $\eta_k \equiv 1$ on $\overline{\Omega}$ and $\eta_k \to 1_{\overline{\Omega}}$ as $k \to \infty$. Let $\varkappa_k := \eta_k \varkappa$. Observing

$$\int_{\hat{\Omega}} 1_{\overline{\Omega}} \varkappa_k : \nabla \hat{w} = \int_{\hat{\Omega}} 1_{\overline{\Omega}} \varkappa : \nabla \hat{w} \,,$$

$$\int_{\hat{\Omega}} f^*(\varkappa_k) \, dx \overset{k \to \infty}{\to} \int_\Omega f^*(\varkappa) \, dx$$

(note: $0 \le f^*(\varkappa_k) \le \eta_k f^*(\varkappa)$, $\varkappa_k \to 1_{\overline{\Omega}} \varkappa$ as $k \to \infty$), we get

$$\int_{\hat{\Omega}} 1_{\overline{\Omega}} \varkappa : \nabla \hat{w} - \int_\Omega f^*(\varkappa) \, dx \le \int_{\hat{\Omega}} 1_{\overline{\Omega}} \varkappa_k : \nabla \hat{w} - \int_{\hat{\Omega}} f^*(\varkappa_k) \, dx + \varepsilon$$

for $k \gg 1$, and the suprema in (12) coincide. This leads to the final representation formula

$$\tilde{J}[w] = \sup \left\{ \int_{\hat{\Omega}} 1_{\overline{\Omega}} \varkappa : \nabla \hat{w} - \int_{\hat{\Omega}} f^*(\varkappa) \, dx : \ \varkappa \in C_0^0 (\hat{\Omega}; \mathbb{R}^{nN}), \right.$$

$$\left. f^* \circ \varkappa \in L^1(\hat{\Omega}) \right\} \tag{13}$$

being valid for any $w \in BV(\Omega; \mathbb{R}^N)$. Now the right-hand side of (13) can be identified with the help of [DT], Proposition 1.2: the right-hand side equals

$$\int_{\overline{\Omega}} f(\nabla^a \hat{w}) \, dx + \int_{\overline{\Omega}} f_\infty \left(\frac{\nabla^s \hat{w}}{|\nabla^s \hat{w}|} \right) d|\nabla^s \hat{w}| \, ,$$

the first integral being equal to $\int_\Omega f(\nabla^a w) \, dx$ and the second integral may be decomposed as $\int_\Omega \cdots + \int_{\partial\Omega} \cdots$, the boundary integral given by (see [AFP], Theorem 3.77, p. 171)

$$\int_{\partial\Omega} f_\infty \left((u_0 - w) \otimes \nu \right) d\mathcal{H}^{n-1} \, ,$$

hence $\tilde{J}[w] = K[w]$ for all $w \in BV(\Omega; \mathbb{R}^N)$. $\square$

A.3 Two uniqueness results

Finally we give two uniqueness results for generalized minimizers. In order to prove the first one, in which we derive a sufficient condition for the uniqueness (up to a constant) of generalized minimizers $u \in \mathcal{M}$, we work with the relaxed functional $\tilde{J}$. Note that the uniqueness up to a constant is what we expect on account of the known results in the minimal surface case (compare the monograph of Giusti, [Giu2]). The idea to establish this first theorem is taken from [Se6] (compare [BF3]): if σ denotes the solution of the dual variational problem as introduced in Section 2.1.1, then for any $u \in \mathcal{M}$ the pair (u, σ) is shown to satisfy an appropriate minimax inequality.

The second theorem is a consequence of the representation formula for $\hat{J}$. Here we have to introduce the subset

$$\mathcal{M}' := \left\{ u \in \mathcal{M} : u \in W^1_{1,loc}(\Omega; \mathbb{R}^N) \right\} = \mathcal{M} \cap W^1_1(\Omega; \mathbb{R}^N) \, .$$

In Theorem A.11 a clear interpretation of the set $\mathcal{M}'$ will be given. To this purpose consider the variational problem

$$\int_\Omega f(\nabla w) \, dx + \int_{\partial\Omega} f_\infty \left((u_0 - w) \otimes \nu \right) d\mathcal{H}^{n-1} \to \min \text{ in } W^1_1(\Omega; \mathbb{R}^N) \, . \quad (\mathcal{P}')$$

It turns out that the set $\mathcal{M}'$ precisely gives the solutions of the problem $(\mathcal{P}')$ and that these solutions are unique up to a constant. This result is of particular interest with respect to Sections 4.2 and 4.3, where under quite general assumptions it is proved that $\mathcal{M}' \neq \emptyset$.

Let us start with

Theorem A.9. *Assume that the variational integrand* $f : \mathbb{R}^{nN} \to [0, \infty)$, $f(0) = 0$, *satisfies Assumption 2.1, in particular the solution* σ *of the dual variational problem* $(\mathcal{P}^*)$ *associated to the problem* $(\mathcal{P})$ *under consideration* *is unique. If σ is a continuous function taking values in $\operatorname{Im}(\nabla f)$, then any generalized minimizer $u \in \mathcal{M}$ satisfies*

$$\nabla u = \nabla f^*(\sigma) \,.$$

Remark A.10.

i) *Of course Theorem A.9 implies the uniqueness of generalized minimizers up to a constant.*

ii) *In particular, the assumptions of Theorem A.9 can be verified for μ-elliptic integrands (with appropriate choice of μ, compare Chapter 4).*

Proof of Theorem A.9. Fix any $u \in \mathcal{M}$ and consider a J-minimizing sequence $\{u_m\}$ in $u_0 + \overset{\circ}{W}{}^1_1(\Omega; \mathbb{R}^N)$ such that

$$u_m \overset{L^{n/(n-1)}}{\rightharpoonup} u \,, \quad u_m \overset{L^1}{\to} u \,.$$

Recalling the definition of $\tilde{l}, \mathcal{U}$ (introduced at the beginning of Appendix A.2) and the representation formula (3), Section 2.1.1, we obtain for any $\varkappa \in \mathcal{U}$

$$J[u_m] = \sup_{\tau \in L^\infty(\Omega; \mathbb{R}^{nN})} l(u_m, \tau) \geq l(u_m, \varkappa) = \tilde{l}(u_m, \varkappa).$$

This gives by passing to the limit $m \to \infty$

$$\inf_{w \in u_0 + \overset{\circ}{W}{}^1_1(\Omega; \mathbb{R}^N)} J[w] \geq \sup_{\varkappa \in \mathcal{U}} \tilde{l}(u, \varkappa) \,.$$

On the other hand, given $\varkappa \in \mathcal{U}$, $v \in u_0 + \overset{\circ}{W}{}^1_1(\Omega; \mathbb{R}^N)$, we observe (recalling that $\operatorname{div} \sigma = 0$)

$$\tilde{l}(u, \varkappa) \leq \inf_{w \in u_0 + \overset{\circ}{W}{}^1_1(\Omega; \mathbb{R}^N)} J[w] = R[\sigma] = \inf_{w \in u_0 + \overset{\circ}{W}{}^1_1(\Omega; \mathbb{R}^N)} l(w, \sigma)$$

$$\leq l(v, \sigma) = \tilde{l}(v, \sigma) = \int_\Omega \sigma : \nabla u_0 \, dx - \int_\Omega f^*(\sigma) \, dx =: \tilde{l}(\sigma) \,.$$

Here, of course, the well known inf-sup relation (see (4), Section 2.1.1) also is exploited. Thus, for any $u \in \mathcal{M}$ and $\varkappa \in \mathcal{U}$ it is proved that

$$\tilde{l}(u, \varkappa) \leq \tilde{l}(\sigma) \,. \tag{14}$$

To proceed further, fix $\lambda \in C^\infty_0(\Omega; \mathbb{R}^{nN})$. By assumption, σ is a continuous function taking values in $\operatorname{Im}(\nabla f)$. Hence, there is a real number $\gamma > 0$ such that $\operatorname{dist}(\sigma(x), \partial \operatorname{Im}(\nabla f)) > \gamma$ for any $x \in \operatorname{spt} \lambda$. If $|t|$ is sufficiently small, then the same is true if we replace σ by $\sigma_t := \sigma + t\lambda$ and γ by $\gamma/2$. Then (14) implies

$$\int_{\text{spt}\,\lambda} \operatorname{div}\sigma_t\,(u_0 - u)\,\mathrm{d}x + \int_{\text{spt}\,\lambda} \sigma_t : \nabla u_0\,\mathrm{d}x$$

$$\leq \int_{\text{spt}\,\lambda} \big(f^*(\sigma_t) - f^*(\sigma)\big)\,\mathrm{d}x + \int_{\text{spt}\,\lambda} \sigma : \nabla u_0\,\mathrm{d}x\,.$$

If we observe that

$$\int_{\text{spt}\,\lambda} (\operatorname{div}\sigma_t)\,u_0\,\mathrm{d}x + \int_{\text{spt}\,\lambda} \sigma_t : \nabla u_0\,\mathrm{d}x - \int_{\text{spt}\,\lambda} \sigma : \nabla u_0\,\mathrm{d}x$$

$$= \int_{\text{spt}\,\lambda} t\,(\operatorname{div}\lambda)\,u_0\,\mathrm{d}x + \int_{\text{spt}\,\lambda} t\lambda : \nabla u_0\,\mathrm{d}x = 0\,,$$

then we obtain

$$-\int_{\text{spt}\,\lambda} t\,(\operatorname{div}\lambda)\,u\,\mathrm{d}x \leq \int_{\text{spt}\,\lambda} \big(f^*(\sigma_t) - f^*(\sigma)\big)\,\mathrm{d}x\,.$$

Dividing through $t > 0$ and passing to the limit $t \to 0$ we get

$$-\int_{\text{spt}\,\lambda} (\operatorname{div}\lambda)\,u\,\mathrm{d}x \leq \int_{\text{spt}\,\lambda} \nabla f^*(\sigma) : \lambda\,\mathrm{d}x\,,$$

i.e., by definition, the first weak derivative of u is given by $\nabla f^*(\sigma)$. $\square$
Now we are going to prove

Theorem A.11. *Suppose that the variational integrand f satisfies Assumption A.1 and assume that there exists $u^* \in \mathcal{M}'$. Then we have*

i) The elements of $\mathcal{M}'$ are solutions of problem $(\mathcal{P}')$ and vice versa.

ii) The set $\mathcal{M}'$ is uniquely determined up to constants.

Proof. ad *i*). On account of the K-minimizing property of $u^* \in \mathcal{M}'$ and since $\nabla^s u^* \equiv 0$, the representation of K clearly implies that $u^* \in \mathcal{M}'$ is a solution of $(\mathcal{P}')$. Conversely, consider a solution v^* of the problem $(\mathcal{P}')$ and a J-minimizing sequence $\{u_m\}$ from $u_0 + \overset{\circ}{W}{}^1_1(\Omega;\mathbb{R}^N)$. The minimality of v^* gives

$$K[v^*] = \int_\Omega f(\nabla v^*)\,\mathrm{d}x + \int_{\partial\Omega} f_\infty\big((u_0 - v^*)\otimes\nu\big)\,\mathrm{d}\mathcal{H}^{n-1} \leq \int_\Omega f(\nabla u_m)\,\mathrm{d}x\,,$$

and *i*) follows from Theorem A.3, *ii*) & *iii*).

ad *ii*). To prove the uniqueness up to a constant, we just observe that f_∞ is convex, whereas f is strictly convex. This immediately gives $\nabla u^* = \nabla u^{**}$ almost everywhere for any two generalized minimizers $u^*, u^{**} \in \mathcal{M}'$, hence Theorem A.11. $\square$

B

Some density results

B.1 Approximations in BV

The standard approximation procedure for functions with bounded variation is given in [AG1] (we refer to [Giu2] Theorem 1.17, p. 14, in particular we also like to mention Remark 2.12, [Giu2], p. 38, on the traces of the approximating sequence), where a sequence of smooth, L^1-converging functions is constructed such that in addition the total variations converge as well. In order to obtain the continuity of the relaxed functional $\hat{J}$ (compare Proposition 2.22) we need a slight modification which, on one hand, is well known (see [AG2], Proposition 2.3). On the other hand, a rigorous proof is hardly found in the literature. Hence, for the sake of completeness, we first show

Lemma B.1. *Let $w \in BV(\Omega; \mathbb{R}^N)$. Then there is a sequence $\{w_m\}$ in $C^\infty(\Omega; \mathbb{R}^N)$ satisfying*

$$\lim_{m \to \infty} \int_\Omega |w_m - w| \, dx \quad = 0 \, ;$$

$$\lim_{m \to \infty} \int_\Omega \sqrt{1 + |\nabla w_m|^2} \, dx = \int_\Omega \sqrt{1 + |\nabla w|^2} \, .$$

Moreover, the trace of each w_m on $\partial\Omega$ coincides with the trace of w.

Proof. To introduce a precise notation which is compatible with [DT] and [Te], respectively, we let

$$g(Z) = \sqrt{1 + |Z|^2} - 1 \quad \text{on } \mathbb{R}^{nN} \, , \tag{1}$$

the conjugate function is given for any $Q \in \mathbb{R}^{nN}$ by

$$g^*(Q) = \begin{cases} +\infty & \text{if } |Q| > 1 \, , \\ 1 - \sqrt{1 - |Q|^2} & \text{if } |Q| \leq 1 \, . \end{cases} \tag{2}$$

Then, for any bounded open Lipschitz domain $U \subset \mathbb{R}^n$ and for any $w \in BV(U; \mathbb{R}^N)$, we take as a definition (compare [DT], Proposition 1.2)

$$\int_U g(\nabla w) := \sup_{\varkappa \in C_0^\infty(U; \mathbb{R}^{nN}), |\varkappa| \le 1} \left\{ -\int_U w \operatorname{div} \varkappa \, dx - \int_U g^*(\varkappa) \, dx \right\} . \qquad (3)$$

Now we follow [Giu2], pp. 14, fix $\varepsilon > 0$ and $w \in BV(\Omega; \mathbb{R}^N)$. Moreover, for any $l \in \mathbb{N}$ we let

$$\Omega_k = \Omega_k^l := \left\{ x \in \Omega : \operatorname{dist}(x, \partial\Omega) > \frac{1}{l+k} \right\}, \qquad k = 0, 1, 2, \dots .$$

Here l may be chosen sufficiently large such that

$$\int_{\Omega - \Omega_0} |\nabla w| < \varepsilon . \qquad (4)$$

Given Ω_k as above, sets A_i are defined by induction: $A_1 := \Omega_2$ and

$$A_i := \Omega_{i+1} - \overline{\Omega}_{i-1} , \qquad i = 2, 3, \dots .$$

With respect to these A_i we then consider a partition of the unity $\{\varphi_i\}$, i.e. for any $i \in \mathbb{N}$

$$\varphi_i \in C_0^\infty(A_i) , \qquad 0 \le \varphi_i \le 1 , \qquad \sum_{i=1}^\infty \varphi_i = 1 .$$

Finally, η denotes a smoothing kernel and we choose for any $i \in \mathbb{N}$ a positive number ε_i sufficiently small such that (letting $\Omega_{-1} = \emptyset$)

$$\operatorname{spt} \eta_{\varepsilon_i} * (\varphi_i w) \subset \Omega_{i+2} - \overline{\Omega}_{i-2} ;$$

$$\int_\Omega |\eta_{\varepsilon_i} * (\varphi_i w) - \varphi_i w| \, dx < 2^{-i}\varepsilon ; \qquad (5)$$

$$\int_\Omega |\eta_{\varepsilon_i} * (w \otimes \nabla\varphi_i) - w \otimes \nabla\varphi_i| \, dx < 2^{-i}\varepsilon .$$

Now assume that $\varepsilon = \frac{1}{m}$ and let

$$w_m = \sum_{i=1}^\infty \eta_{\varepsilon_i} * (\varphi_i w) .$$

Note that here and in the following each sum under consideration is locally finite.

The first claim of the lemma is immediate since $\{w_m\}$ is constructed as a smooth sequence and since we have recalling (5)

$$\int_\Omega |w_m - w| \, dx \le \sum_{i=1}^\infty \int_\Omega |\eta_{\varepsilon_i} * (\varphi_i w) - \varphi_i w| \, dx \le \varepsilon .$$

This, together with lower semicontinuity, also proves

$$\int_\Omega g(\nabla w) \le \liminf_{m\to\infty} \int_\Omega g(\nabla w_m)\, dx \ . \tag{6}$$

To establish the opposite inequality we fix $\varkappa \in C_0^\infty(\Omega; \mathbb{R}^{nN})$, $|\varkappa| \le 1$. Then, by (4), (5), on account of $\sum_{i=1}^\infty \nabla\varphi_i \equiv 0$ and since the intersection of more than three A_i is empty, we obtain

$$- \int_\Omega w_m \operatorname{div} \varkappa \, dx - \int_\Omega g^*(\varkappa)\, dx$$

$$= - \int_\Omega w \operatorname{div}\left(\varphi_1\, \eta_{\varepsilon_1} * \varkappa\right) dx - \sum_{i=2}^\infty \int_\Omega w \operatorname{div}\left(\varphi_i\, \eta_{\varepsilon_i} * \varkappa\right) dx$$

$$+ \sum_{i=1}^\infty \int_\Omega \varkappa : \left(\eta_{\varepsilon_i} * (w \otimes \nabla\varphi_i) - w \otimes \nabla\varphi_i\right) dx - \int_\Omega g^*(\varkappa)\, dx \tag{7}$$

$$\le - \int_\Omega w \operatorname{div}\left(\varphi_1\, \eta_{\varepsilon_1} * \varkappa\right) dx - \int_\Omega g^*(\varkappa)\, dx + 4\varepsilon \ .$$

If we identify $\varkappa$ with its zero-extension to $\mathbb{R}^n$, then we may write

$$- \int_\Omega g^*(\varkappa)\, dx = - \int_{\mathbb{R}^n} g^*(\varkappa)\, dx$$

$$= - \int_{\mathbb{R}^n} g^*(\eta_{\varepsilon_1} * \varkappa)\, dx + \left\{ \int_{\mathbb{R}^n} \left[g^*(\eta_{\varepsilon_1} * \varkappa) - g^*(\varkappa) \right] dx \right\}$$

$$= I + II \ . \tag{8}$$

An estimate for I follows from $0 \le \varphi_1 \le 1$, the convexity of g^*, $g^*(0) = 0$ and $g^* \ge 0$:

$$I \le - \int_{\mathbb{R}^n} g^*(\varphi_1\, \eta_{\varepsilon_1} * \varkappa)\, dx = - \int_\Omega g^*(\varphi_1\, \eta_{\varepsilon_1} * \varkappa)\, dx \ . \tag{9}$$

To handle the second integral of (8) we use Jensen's inequality

$$g^*(\eta_{\varepsilon_1} * \varkappa) \le \eta_{\varepsilon_1} * g^*(\varkappa) \ .$$

This, together with standard calculations, gives

$$II \le \int_{\mathbb{R}^n} \left[\eta_{\varepsilon_1} * g^*(\varkappa) - g^*(\varkappa) \right] dx = 0 \ . \tag{10}$$

As a result of (7)–(10) it is proved that

$$- \int_\Omega w_m \operatorname{div} \varkappa \, dx - \int_\Omega g^*(\varkappa)\, dx$$

$$\le - \int_\Omega w \operatorname{div}\left(\varphi_1\, \eta_{\varepsilon_1} * \varkappa\right) dx - \int_\Omega g^*(\varphi_1\, \eta_{\varepsilon_1} * \varkappa)\, dx + 4\varepsilon \ . \tag{11}$$

Now it is obvious that the comparison function on the right-hand side of (11) is admissible and the opposite inequality to (6) is established:

$$\limsup_{m \to \infty} \int_\Omega g(\nabla w_m) \, dx \le \int_\Omega g(\nabla w) \, .$$

Moreover, the approximative functions w_m are defined in the same manner as in [Giu2], pp. 14, hence Remark 2.12 of [Giu2], p. 38, remains valid which proves that the trace of each w_m on $\partial\Omega$ coincides with the trace of w.   □

As a generalization of Lemma B.1 we now choose an approximative sequence with arbitrary prescribed traces. To this purpose we fix again a bounded Lipschitz domain $\hat{\Omega} \supseteq \Omega$ and extend our boundary values $u_0 \in W_1^1(\Omega; \mathbb{R}^N)$ to a function of class $\overset{\circ}{W}_1^1(\hat{\Omega}; \mathbb{R}^N)$.

Lemma B.2. *Given $\hat{\Omega}$, u_0 as above let $w \in BV(\Omega; \mathbb{R}^N)$ and denote by $\hat{w}$ the extension via u_0 to the domain $\hat{\Omega}$. Then there exists a sequence $\{w_m\}$ in $u_{0|\Omega} + C_0^\infty(\Omega; \mathbb{R}^N)$ such that if $m \to \infty$ (and if we extend w_m by u_0 to $\hat{\Omega}$)*

i) $\qquad w_m \to \hat{w} \quad in \; L^1(\hat{\Omega}; \mathbb{R}^N);$

ii) $\qquad \displaystyle\int_{\hat{\Omega}} \sqrt{1 + |\nabla w_m|^2} \, dx \to \int_{\hat{\Omega}} \sqrt{1 + |\nabla \hat{w}|^2}.$

Proof. Let us recall the general setting (1)–(3) and fix w, $\hat{w}$ as given in the lemma. Now we use a construction as outlined in [Alt], p. 170. Since $\partial\Omega$ is Lipschitz, we can cover $\partial\Omega$ by open sets $V_1, \ldots, V_r$ such that after rotation and translation V_j takes the form

$$V_j = \left\{ x \in \mathbb{R}^n : \left|(x_1, \ldots, x_{n-1})\right| < r_j, \; |x_n - g_j(x_1, \ldots, x_{n-1})| < h_j \right\},$$

where g_j is a Lipschitz function. Moreover, we have

$$0 = x_n - g_j(x_1, \ldots, x_{n-1}) \qquad \Rightarrow x \in \partial\Omega \, ;$$

$$0 < x_n - g_j(x_1, \ldots, x_{n-1}) < h_j \quad \Rightarrow x \in \Omega \, ;$$

$$0 > x_n - g_j(x_1, \ldots, x_{n-1}) > -h_j \Rightarrow x \notin \Omega \, .$$

Let V_{r+1}, V_{r+2} denote open sets such that $\overline{V}_{r+1} \subset \Omega$, $\overline{V}_{r+2} \subset \mathbb{R}^n - \overline{\Omega}$ and

$$\hat{\Omega} \subset \bigcup_{j=1}^{r+2} V_j \, .$$

In addition, the sets V_j may be chosen such that for any $j = 1, \ldots, r+2$

$$g(\nabla \hat{w})(\partial V_j) = 0 \, .$$

Here we let for any Borel set $B \subset \hat{\Omega}$ (again following [DT], see also the proof of Theorem A.6)

$$g(\nabla \hat{w})(B) = \int_B g(\nabla \hat{w})$$

$$= \sup_{\varkappa \in C_0^\infty(\hat{\Omega};\mathbb{R}^{nN}),\, |\varkappa| \leq 1} \left\{ \int_{\hat{\Omega}} 1_B \varkappa \nabla \hat{w} - \int_{\hat{\Omega}} g^*(\varkappa)\, dx \right\}.$$

Now fix $\varepsilon > 0$ and decrease V_j if necessary (to be a little bit more precise: decrease radius and height of V_j a little bit and obtain sets $V_{j,\varepsilon} \Subset V_j$. Then let $U_1 = V_1$, $U_2 = V_2 - \overline{V}_{1,\varepsilon}$ and inductively define U_j, $j = 1,\ldots,r+2$.). As a result we obtain smooth sets U_j which can be arranged in addition such that

$$\sum_{j=1}^{r+2} g(\nabla \hat{w})(U_j) \leq g(\nabla \hat{w})(\hat{\Omega}) + \varepsilon, \tag{12}$$

$$g(\nabla \hat{w})(\partial U_j) = 0, \quad j = 1,\ldots,r+2. \tag{13}$$

Note that $u_0 \in \overset{\circ}{W}{}_1^1(\hat{\Omega};\mathbb{R}^N)$, hence we may extend $\nabla \hat{w}$ by 0 to $\mathbb{R}^n - \hat{\Omega}$. Subordinate to the covering $\{U_j\}$ let $\{\varphi_j\}$ denote a partition of the unity, i.e.

$$\varphi_j \in C_0^\infty(U_j), \quad 0 \leq \varphi_j \leq 1, \quad \sum_{j=0}^{r+2} \varphi_j \equiv 1 \text{ on } \hat{\Omega}.$$

For each fixed index $1 \leq j \leq r$ and for $\delta \ll 1$ we then define (neglecting with a slight abuse of notation rotations and translations)

$$\hat{v}_j^{\delta,\varepsilon}(x) := \varphi_j(x) \left[(\hat{w} - u_0) \circ (x - \delta e_n) + u_0(x) \right].$$

Finally, we let $\hat{v}_{r+1}^{\delta,\varepsilon} := \varphi_{r+1} w$, $\hat{v}_{r+2}^{\delta,\varepsilon} := \varphi_{r+2} u_0$ and

$$\hat{v}^{\delta,\varepsilon} := \sum_{j=1}^{r+2} \hat{v}_j^{\delta,\varepsilon}.$$

Clearly $\hat{v}^{\delta,\varepsilon} - u_0$ is compactly supported in Ω. Moreover, for $\delta = \delta(\varepsilon)$ sufficiently small we have

$$\int_{\hat{\Omega}} \left| \hat{v}^{\delta(\varepsilon),\varepsilon} - \hat{w} \right| dx \leq \varepsilon,$$

and again lower semicontinuity gives for $\hat{v}^\varepsilon = \hat{v}^{\delta(\varepsilon),\varepsilon}$

$$\int_{\hat{\Omega}} g(\nabla \hat{w}) \leq \liminf_{\varepsilon \to 0} \int_{\hat{\Omega}} g(\nabla \hat{v}^\varepsilon)\, dx. \tag{14}$$

To prove the opposite inequality we recall the definition (3), fix $\varkappa \in C_0^\infty(\hat{\Omega};\mathbb{R}^{nN})$, $|\varkappa| \leq 1$, and observe

$$-\int_{\hat{\Omega}} \hat{v}^{\varepsilon}\,\mathrm{div}\,\varkappa\,\mathrm{d}x = -\sum_{j=1}^{r}\int_{U_j}\varphi_j(x)\,\hat{w}\big(x-\delta(\varepsilon)\,e_n\big)\,\mathrm{div}\,\varkappa\,\mathrm{d}x$$

$$-\sum_{j=r+1}^{r+2}\int_{U_j}\varphi_j(x)\,\hat{w}(x)\,\mathrm{div}\,\varkappa\,\mathrm{d}x$$

$$+\sum_{j=1}^{r}\int_{U_j}\varkappa:\nabla\Big[\varphi_j\big\{u_0(x)-u_0\big(x-\delta(\varepsilon)\,e_n\big)\big\}\Big]\,\mathrm{d}x$$

$$\leq -\sum_{j=1}^{r}\int_{U_j}\varphi_j(x)\,\hat{w}\big(x-\delta(\varepsilon)\,e_n\big)\,\mathrm{div}\,\varkappa\,\mathrm{d}x$$

$$-\sum_{j=r+1}^{r+2}\int_{U_j}\varphi_j(x)\,\hat{w}(x)\,\mathrm{div}\,\varkappa\,\mathrm{d}x+\varepsilon\,,$$

where $\delta(\varepsilon)$ is assumed to be sufficiently small. Note that this choice does not depend on $\varkappa$. Now, the right-hand side is estimated by

$$\mathrm{r.h.s.} = -\sum_{j=1}^{r}\int_{U_j}\hat{w}\big(x-\delta(\varepsilon)\,e_n\big)\,\mathrm{div}\,(\varphi_j\,\varkappa)\,\mathrm{d}x$$

$$+\sum_{j=1}^{r}\int_{U_j}\hat{w}\big(x-\delta(\varepsilon)\,e_n\big)\otimes\nabla\varphi_j:\varkappa\,\mathrm{d}x$$

$$-\sum_{j=r+1}^{r+2}\int_{U_j}\hat{w}(x)\,\mathrm{div}\,(\varphi_j\,\varkappa)\,\mathrm{d}x + \sum_{j=r+1}^{r+2}\int_{U_j}\hat{w}(x)\otimes\nabla\varphi_j:\varkappa\,\mathrm{d}x+\varepsilon$$

$$\leq -\sum_{j=1}^{r}\int_{U_j}\hat{w}\big(x-\delta(\varepsilon)\,e_n\big)\,\mathrm{div}\,(\varphi_j\,\varkappa)\,\mathrm{d}x$$

$$-\sum_{j=r+1}^{r+2}\int_{U_j}\hat{w}(x)\,\mathrm{div}\,(\varphi_j\,\varkappa)\,\mathrm{d}x$$

$$+\sum_{j=1}^{r+2}\int_{U_j}\hat{w}(x)\otimes\nabla\varphi_j:\varkappa\,\mathrm{d}x$$

$$+\sum_{j=1}^{r}\int_{U_j}\big|\hat{w}(x)-\hat{w}\big(x-\delta(\varepsilon)\,e_n\big)\big|\,|\nabla\varphi_j|\,\mathrm{d}x+\varepsilon\,.$$

Once more $\delta(\varepsilon)$ is decreased in order to bound the last sum by ε. Moreover, $\sum_{j=1}^{r+2}\nabla\varphi_j\equiv 0$ implies the third sum to vanish identically.

Next observe that $g^*(\varphi_j\,\varkappa)\leq\varphi_j\,g^*(\varkappa)$ follows from the convexity of g^* and from $g^*(0)=0$, thus we arrive at ($\varkappa\equiv 0$ on $\mathbb{R}^n-\hat{\Omega}$)

$$-\int_{\hat{\Omega}} \hat{v}^\varepsilon \operatorname{div} \varkappa \, dx - \int_{\hat{\Omega}} g^*(\varkappa) \, dx$$

$$\leq \sum_{j=1}^{r} \left\{ -\int_{U_j} \hat{w}\left(x - \delta(\varepsilon) e_n\right) \operatorname{div}\left(\varphi_j \varkappa\right) dx - \int_{U_j} g^*\left(\varphi_j \varkappa\right) dx \right\} \qquad (15)$$

$$+ \sum_{i=r+1}^{r+2} \left\{ -\int_{U_j} \hat{w}(x) \operatorname{div}\left(\varphi_j \varkappa\right) dx - \int_{U_j} g^*\left(\varphi_j \varkappa\right) dx \right\} + 2\varepsilon \, .$$

Finally, we choose open sets $\tilde{U}_j \ni U_j$, $j = 1, \ldots, r$ such that

$$\sum_{j=1}^{r} g(\nabla \hat{w})(\tilde{U}_j) \leq \sum_{j=1}^{r} g(\nabla \hat{w})(U_j) + \varepsilon \, . \qquad (16)$$

Note that sets of this kind can be found on account of (13). Moreover, if $\delta(\varepsilon) \ll 1$, then we remark that for $j = 1, \ldots, r$

$$-\int_{U_j} \hat{w}\left(x - \delta(\varepsilon) e_n\right) \operatorname{div}\left(\varphi_j \varkappa\right) dx - \int_{U_j} g^*\left(\varphi_j \varkappa\right) dx$$

$$\leq \sup_{\tilde{\varkappa} \in C_0^\infty(\tilde{U}_j; \mathbb{R}^{nN}), \, |\tilde{\varkappa}| \leq 1} \left\{ -\int_{\tilde{U}_j} \hat{w} \operatorname{div} \tilde{\varkappa} \, dx - \int_{\tilde{U}_j} g^*(\tilde{\varkappa}) \, dx \right\} \, . \qquad (17)$$

Putting together the inequalities (15)–(17) it is proved that

$$-\int_{\hat{\Omega}} \hat{v}^\varepsilon \operatorname{div} \varkappa \, dx - \int_{\hat{\Omega}} g^*(\varkappa) \, dx$$

$$\leq \sum_{j=1}^{r} g(\nabla \hat{w})(\tilde{U}_j) + \sum_{j=r+1}^{r+2} g(\nabla \hat{w})(U_j) + 2\varepsilon \qquad (18)$$

$$\leq \sum_{j=r+1}^{r+2} g(\nabla \hat{w})(U_j) + 3\varepsilon \, .$$

Once (18) is established, (12) shows the opposite inequality to (14). Summarizing the results we have proved up to now: there exists a sequence $\{\hat{v}^\varepsilon\}$ such that $\hat{v}^\varepsilon - u_0$ is compactly supported in Ω and such that the convergences claimed in the lemma hold for this sequence. In a last step it remains to apply the standard smoothing procedure (see Lemma B.1) and Lemma B.2 is proved. $\square$

B.2 A density result for $\mathcal{U} \cap L(c)$

Here we are going to establish a density result from [BF4] which was needed for the proof of the identification Theorem A.6. With the notation $\mathcal{U}$ and $L(c)$ as introduced in Appendix A.2, a precise formulation of this lemma reads as

Lemma B.3. *Suppose that $\varkappa \in \mathcal{U}$ satisfies $\varkappa(x) \in L(c)$ for some $c \in \mathbb{R}$. Then a sequence $\{\varkappa_m\}$, $\varkappa_m \in C^\infty(\overline{\Omega}; \mathbb{R}^{nN})$ exists such that*

i) $\varkappa_m \to \varkappa$ *in* $L^t(\Omega; \mathbb{R}^{nN})$ *for any $t < \infty$ and we have almost everywhere convergence;*

ii) $\operatorname{div} \varkappa_m \to \operatorname{div} \varkappa$ *in* $L^n(\Omega; \mathbb{R}^N)$;

iii) $\varkappa_m \stackrel{*}{\to} \varkappa$ *in* $L^\infty(\Omega; \mathbb{R}^{nN})$;

iv) $\varkappa_m(x) \in L(c)$ *for all $x \in \overline{\Omega}$ and for any $m \in \mathbb{N}$.*

Proof. As in the proof of Lemma B.2, the boundary of the Lipschitz domain Ω is covered with sets V_j, $j = 1, \ldots, r$, such that we have the properties stated there. Let V_0 denote an open set satisfying $\overline{V}_0 \subset \Omega$,

$$\overline{\Omega} \subset \bigcup_{j=0}^r V_j \, ,$$

and consider a corresponding partition of the unity $\{\varphi_j\}$, i.e.

$$\varphi_j \in C_0^\infty(V_j) \, , \quad 0 \le \varphi_j \le 1 \, , \quad \sum_{j=0}^r \varphi_j \equiv 1 \ \text{on} \ \overline{\Omega} \, .$$

For each fixed index $j \ge 1$ we let for $\delta \ll 1$

$$\varkappa_j^\delta(x) := \begin{cases} \varkappa(x + \delta e_n)\varphi_j(x) & \text{if } x \in \Omega \cap V_j \, ; \\ 0 & \text{if } x \in \Omega - V_j \, . \end{cases}$$

Note that $\varkappa_j^\delta \equiv 0$ near the "upper" boundary part of V_j, the same is true near the "vertical" boundary parts which follows from the support properties of φ_j and from an appropriate choice of δ. If ω denotes a smoothing kernel, we let

$$\varkappa^{\delta,\rho}(x) := \omega_\rho * \left[\varphi_0 \varkappa + \sum_{j=1}^r \varkappa_j^\delta\right](x) \, , \quad x \in \overline{\Omega} \, .$$

Assuming again the standard representation of the neighborhood V_j we get

$$\omega_\rho * \varkappa_j^\delta(x) = \int_{\mathbb{R}^n} \omega_\rho(y - x)\varkappa(y + \delta e_n)\varphi_j(y) \, \mathrm{d}y \, ,$$

and for ρ small enough depending on δ we see that for $y \in B_\rho(x)$, $x \in \overline{\Omega}$, the point $y + \delta e_n$ belongs to Ω, and $\omega_\rho * \varkappa_j^\delta(x)$ is well defined. Clearly $\omega_\rho * \varkappa_j^\delta \in C^\infty(\overline{\Omega}; \mathbb{R}^{nN})$ and

$$\omega_\rho * \varkappa_j^\delta \stackrel{\rho\downarrow 0}{\to} \varkappa_j^\delta \quad \text{in} \ L^p(\Omega; \mathbb{R}^{nN})$$

for any $p < \infty$, moreover (see [Alt], Lemma 1.16, p. 18)

$$\varkappa_j^\delta \overset{\delta \downarrow 0}{\to} \varkappa \varphi_j \quad \text{in } L^p(\Omega; \mathbb{R}^{nN}),$$

again for any $p < \infty$. We further have for $x \in \Omega$

$$\operatorname{div}\left(\omega_\rho * \varkappa_j^\delta\right)(x) = -\int_{B_\rho(x)} (\partial_\alpha \omega_\rho)(y - x)\,\varkappa_\alpha(y + \delta e_n)\,\varphi_j(y)\,dy$$

and since it is sufficient to consider $x \in \Omega \cap V_j$, we see that the arguments on the right-hand side are compactly supported in Ω. Moreover, $y \mapsto \omega_\rho(y - x)$ has compact support in $B_\rho(x)$, thus

$$\operatorname{div}\left(\omega_\rho * \varkappa_j^\delta\right)(x) = \int_{B_\rho(x)} \omega_\rho(y - x)\Big[\operatorname{div}\varkappa(y + \delta e_n)\,\varphi_j(y)$$

$$+ \varkappa(y + \delta e_n)\,\nabla\varphi_j(y)\Big]\,dy$$

and, as above,

$$\operatorname{div}\left(\omega_\rho * \varkappa_j^\delta\right) \overset{\rho \downarrow 0}{\to} \operatorname{div}\varkappa(\cdot + \delta e_n)\,\varphi_j + \varkappa(\cdot + \delta e_n)\,\nabla\varphi_j$$

in $L^n(\Omega; \mathbb{R}^N)$. The right-hand side converges to

$$\operatorname{div}\varkappa\,\varphi_j + \varkappa\,\nabla\varphi_j$$

in $L^n(\Omega; \mathbb{R}^N)$ as $\delta \downarrow 0$. So, if we first fix a sequence $\delta_m \downarrow 0$, we find a sequence $\{\rho_m\}$ depending on $\{\delta_m\}$ such that the convergence properties $i)$ and $ii)$ hold for $\varkappa^m := \varkappa^{\delta_m, \rho_m}$. The boundedness of $\|\varkappa^m\|_{L^\infty(\Omega; \mathbb{R}^{nN})}$ implies $\varkappa^m \overset{*}{\to} \tilde{\varkappa}$ in $L^\infty(\Omega; \mathbb{R}^{nN})$ for a subsequence and some tensor $\tilde{\varkappa} \in L^\infty(\Omega; \mathbb{R}^{nN})$, but $i)$ shows $\tilde{\varkappa} = \varkappa$. It remains to prove $iv)$. Jensen's inequality applied to the measure $\omega_\rho(x - \cdot)\mathcal{L}^n$ gives

$$f^*\left(\varkappa^{\delta, \rho}(x)\right) \le \int_{B_\rho(x)} \omega_\rho(x - y)\,f^*\left(\varphi_0 \varkappa + \sum_{j=1}^r \varkappa_j^\delta\right)(y)\,dy$$

and if we recall the definition of $\varkappa_j^\delta$ we see that f^* is evaluated on the convex combination

$$\varphi_0(y)\,\varkappa(y) + \sum_{j=1}^r \varphi_j(y)\,\varkappa(\dots),$$

where $\varkappa(\dots)$ has an obvious meaning for $j = 1, \dots, r$. Our assumption $\varkappa \in L(c)$ almost everywhere then implies

$$f^*\left(\varphi_0 \varkappa + \sum_{j=1}^r \varkappa_j^\delta\right)(y) \le c,$$

i.e. $\varkappa^{\delta,\rho} \in L(c)$. □

For technical reasons (see the proof of Theorem A.6), we also need the following

Remark B.4. *Recall the definition of $\varkappa_j^\delta(x)$ given in the proof of Lemma B.3. Clearly this definition makes sense for points x such that $x + \delta e_n \in \overline{\Omega}$, i.e.*

$$-\delta \leq x_n - g_j(x_1, \ldots, x_{n-1}),$$

so that (combined with the smoothing procedure outlined above) $\varkappa_j^\delta \in C^\infty(V_j \cap [-\delta/2 \leq x_n - g_j(x_1, \ldots, x_{n-1})])$. If we then let

$$\psi_j^\delta(x) := \begin{cases} 1 \text{ if } x \in V_j,\ x_n - g_j(x_1, \ldots, x_{n-1}) \geq 0, \\ 0 \text{ if } x \in V_j,\ x_n - g_j(x_1, \ldots, x_{n-1}) \leq -\delta/4, \end{cases}$$

$\psi_j^\delta \in C_0^\infty(\mathbb{R}^n)$, $0 \leq \psi \leq 1$, then the function $\psi_j^\delta \varkappa_j^\delta$, $j = 1, \ldots, r$, is of class $C_0^\infty(\hat\Omega; \mathbb{R}^{nN})$, where $\hat\Omega \supseteq \Omega$ is some bounded Lipschitz domain and δ is chosen sufficiently small. As a result, if $\hat\Omega$ is fixed as above, then the sequence $\{\varkappa_m\}$ given in Lemma B.3 may be in addition assumed to be of class $C_0^\infty(\hat\Omega; \mathbb{R}^{nN})$. Moreover, again by the convexity of f^* (further recall that $f^*(0) = 0$), the level set property continues to hold on the extended domain $\hat\Omega$.

B.3 Local comparison functions

A helpful tool which was used in Sections 2.3, 2.4 and 4.3 is the construction of local comparison functions as given in [BF1]. With the notation

$$J[w; \hat\Omega] = \int_{\hat\Omega} f(\nabla w)\, dx \quad \text{for any open set } \hat\Omega$$

we now prove for f given as in Assumption 2.1

Lemma B.5. *Consider a sequence $\{u_m\} \subset W_1^1(\Omega; \mathbb{R}^N)$ such that:*

i) $u_m \to u^*$ in $L^1(\Omega; \mathbb{R}^N)$ as $m \to \infty$;

ii) $\sup_{m \in \mathbb{N}} \|u_m\|_{W_1^1(\Omega; \mathbb{R}^N)} < \infty$.

Then we can find a subsequence (still denoted by $\{u_m\}$) with the following properties: for any $x_0 \in \Omega$ and for almost any ball $B_R(x_0)$, $B_{2R}(x_0) \Subset \Omega$, there is a sequence $\{w_m\} \subset W_1^1(\Omega; \mathbb{R}^N)$ such that

i) $w_m \to u^*$ in $L^1(\Omega; \mathbb{R}^N)$ as $m \to \infty$;

ii) $\limsup_{m \to \infty} J[w_m; B_R(x_0)] \leq \liminf_{m \to \infty} J[u_m; B_R(x_0)]$;

iii) $\limsup\limits_{m \to \infty} J[w_m; \Omega] \le \liminf\limits_{m \to \infty} J[u_m; \Omega]$;

iv) $w_{m|\partial B_R(x_0)} = u^*_{|\partial B_R(x_0)}$, *where the traces are well defined functions of class* $L^1(\partial B_R(x_0); \mathbb{R}^N)$;

v) $w_{m|\Omega - B_{2R}(x_0)} = u_{m_k|\Omega - B_{2R}(x_0)}$ *for any* $m \in \mathbb{N}$, *in particular the boundary values on* $\partial\Omega$ *are preserved;*

vi) $w_{m|B_{R/2}(x_0)} = u_{m_l|B_{R/2}(x_0)}$ *for any* $m \in \mathbb{N}$.

Here $\{u_{m_k}\}$ *and* $\{u_{m_l}\}$ *denote some appropriate subsequences of* $\{u_m\}$.

Proof. We have $u^* \in BV$ and we may also assume that

$$\nabla u_m \rightharpoonup \nabla u^*, \quad |\nabla u_m| \rightharpoonup \mu \quad \text{as } m \to \infty$$

in the sense of measures where μ denotes a Radon measure of finite mass. We may choose a radius $R > 0$ according to

$$B_{2R}(x_0) \Subset \Omega \quad \text{and} \quad \mu(\partial B_R(x_0)) = 0 = |\nabla u^*|(\partial B_R(x_0)) . \quad (19)$$

This implies (see [Giu2], Remark 2.13) that $u^*_{|\partial B_R(x_0)}$ is well defined. With $T_\varepsilon := \{x : R - \varepsilon < |x - x_0| < R + \varepsilon\}$, $\varepsilon > 0$ sufficiently small, we further obtain:

$$\lim_{\varepsilon \to 0} \left\{ \limsup_{m \to \infty} \int_{T_\varepsilon} |\nabla u_m| \, dx \right\} = 0 , \quad (20)$$

$$\lim_{\varepsilon \to 0} \int_{T_\varepsilon} |\nabla u^*| = 0 . \quad (21)$$

For (21) we just observe using (19)

$$\int_{T_\varepsilon} |\nabla u^*| \overset{\varepsilon \to 0}{\longrightarrow} |\nabla u^*| \left(\bigcap_{\delta > 0} T_\delta \right) = |\nabla u^*|(\partial B_R(x_0)) = 0 .$$

Next, let $\varphi_\varepsilon \in C_0(T_{2\varepsilon}, [0,1])$, $\varphi_\varepsilon = 1$ on T_ε. Then

$$\int_{T_\varepsilon} |\nabla u_m| \, dx \le \int_{T_{2\varepsilon}} \varphi_\varepsilon |\nabla u_m| \, dx \overset{m \to \infty}{\longrightarrow} \int_{T_{2\varepsilon}} \varphi_\varepsilon \, d\mu ,$$

hence $\quad \limsup\limits_{m \to \infty} \int_{T_\varepsilon} |\nabla u_m| \, dx \le \mu(T_{2\varepsilon}) \overset{\varepsilon \to 0}{\longrightarrow} \mu(\partial B_R(x_0)) = 0 ,$

thus (20) holds. With R fixed we now let

$$\mathbb{K} := \left\{ w \in W_1^1(B_R(x_0); \mathbb{R}^N) : w_{|\partial B_R(x_0)} = u^*_{|B_R(x_0)} \right\} .$$

Note that $\mathbb{K} \neq \emptyset$ on account of [Giu2], Remark 2.12 and [Giu2], Theorem 2.16. We first claim that there exists a sequence $\{v_k\} \subset \mathbb{K}$ for which conclusion (ii.) holds:

$$\limsup_{k \to \infty} J\big[v_k; B_R(x_0)\big] \leq \liminf_{m \to \infty} J\big[u_m; B_R(x_0)\big] . \qquad (22)$$

For proving (22) consider $\tilde{u}_m \in C^\infty(B_R(x_0), \mathbb{R}^N)$ with (compare again [Giu2], Remark 2.12)

$$\int_{B_R(x_0)} |\tilde{u}_m - u^*| \, dx \to 0 \,, \quad \int_{B_R(x_0)} |\nabla \tilde{u}_m| \, dx \to \int_{B_R(x_0)} |\nabla u^*|$$

as $m \to \infty$ and such that $\tilde{u}_{m|\partial B_R(x_0)} = u^*_{|\partial B_R(x_0)}$. Then

$$\lim_{\varepsilon \to 0} \left(\limsup_{m \to \infty} \int_{B_R(x_0) \cap T_\varepsilon} |\nabla \tilde{u}_m| \, dx \right) = 0 . \qquad (23)$$

In fact, let $A_\varepsilon := \{x : R - \varepsilon < |x - x_0| < R\}$. We have (see [Giu2], Prop. 1.13)

$$\limsup_{m \to \infty} \int_{B_R(x_0) \cap T_\varepsilon} |\nabla \tilde{u}_m| \, dx = \limsup_{m \to \infty} \int_{B_R(x_0) \cap A_\varepsilon} |\nabla \tilde{u}_m| \, dx$$

$$\leq \int_{B_R(x_0) \cap \overline{A}_\varepsilon} |\nabla u^*| \leq \int_{T_{2\varepsilon}} |\nabla u^*| \to 0$$

as $\varepsilon \to 0$ and (23) follows. We define

$$v_m^\varepsilon := \begin{cases} u_m & \text{on } B_R(x_0) - T_\varepsilon \,, \\ u_m + \varepsilon^{-1}(\tilde{u}_m - u_m)(|x| - R + \varepsilon) & \text{on } A_\varepsilon = B_R(x_0) \cap T_\varepsilon \,. \end{cases}$$

Both parts of v_m^ε induce the same trace on $\partial B_{R-\varepsilon}(x_0)$, thus v_m^ε is of class $W_1^1(B_R(x_0); \mathbb{R}^N)$ and in addition $v_m^\varepsilon \in \mathbb{K}$. Since f is of linear growth, the behavior of $\int_{A_\varepsilon} f(\nabla v_m^\varepsilon) \, dx$ follows from the discussion of $\int_{A_\varepsilon} |\nabla v_m^\varepsilon| \, dx$:

$$\int_{A_\varepsilon} |\nabla v_m^\varepsilon| \, dx \leq 2 \int_{A_\varepsilon} |\nabla u_m| \, dx + \int_{A_\varepsilon} |\nabla \tilde{u}_m| \, dx + \frac{1}{\varepsilon} \int_{B_R(x_0)} |u_m - \tilde{u}_m| \, dx .$$

According to (20) and (23) it is possible to define a sequence $\varepsilon_k \to 0$ such that

$$\limsup_{m \to \infty} \int_{A_{\varepsilon_k}} |\nabla u_m| \, dx \leq \frac{1}{k} \quad \text{and} \quad \limsup_{m \to \infty} \int_{A_{\varepsilon_k}} |\nabla \tilde{u}_m| \, dx \leq \frac{1}{k}$$

for all $k \in \mathbb{N}$, thus, for any $k \in \mathbb{N}$, there is $m_k \in \mathbb{N}$ such that

$$\int_{A_{\varepsilon_k}} |\nabla u_m| \, dx \leq \frac{2}{k} \quad \text{and} \quad \int_{A_{\varepsilon_k}} |\nabla \tilde{u}_m| \, dx \leq \frac{2}{k}$$

for all $m \geq m_k$. Recalling the L^1–convergences $u_m, \tilde{u}_m \to u^*$ on $B_R(x_0)$, we assume in addition

$$\frac{1}{\varepsilon_k} \int_{B_R(x_0)} |u_m - \tilde{u}_m| \, \mathrm{d}x \leq \varepsilon_k$$

for all $m \geq m_k$, $k \in \mathbb{N}$. Putting together these estimates we get

$$J[v_m^{\varepsilon_k}; B_R(x_0)] \leq \int_{B_R(x_0) - T_{\varepsilon_k}} f(\nabla u_m) \, \mathrm{d}x + \alpha_k , \quad m \geq m_k , \quad k \in \mathbb{N},$$

for a sequence $\alpha_k \geq 0$, $\alpha_k \to 0$ as $k \to \infty$. By enlarging m_k, if necessary, we can arrange

$$\lim_{k \to \infty} \int_{B_R(x_0)} f(\nabla u_{m_k}) \, \mathrm{d}x = \liminf_{l \to \infty} \int_{B_R(x_0)} f(\nabla u_l) \, \mathrm{d}x .$$

Finally $v_k := v_{m_k}^{\varepsilon_k}$ is introduced. Then

$$\limsup_{k \to \infty} J[v_k; B_R(x_0)] \leq \limsup_{k \to \infty} \int_{B_R(x_0)} f(\nabla u_{m_k}) \, \mathrm{d}x ,$$

and (22) is established. From the definition of v_k it also follows that

$$\lim_{k \to \infty} \int_{B_R(x_0)} |v_k - u^*| \, \mathrm{d}x = 0 . \tag{24}$$

Note that by construction we clearly may assume that $v_k|_{B_{R/2}(x_0)} = u_{m_i}|_{B_{R/2}(x_0)}$ for any $k \in \mathbb{N}$ and for some subsequence of $\{u_m\}$.

Next, an analogous construction is needed in the exterior domain: choose some radius $\rho > R$, $B_\rho(x_0) \Subset \Omega$, and a sequence $\hat{u}_m \in C^\infty(B_\rho(x_0) - B_R(x_0); \mathbb{R}^N)$ satisfying

$$\hat{u}_m|_{\partial(B_\rho(x_0) - B_R(x_0))} = u^*_{|\partial(B_\rho(x_0) - B_R(x_0))} ,$$

$$\hat{u}_m \to u^* \text{ in } L^1(B_\rho(x_0) - B_R(x_0); \mathbb{R}^N)$$

and $\displaystyle \int_{B_\rho(x_0) - B_R(x_0)} |\nabla \hat{u}_m| \, \mathrm{d}x \to \int_{B_\rho(x_0) - B_R(x_0)} |\nabla u^*| \quad \text{as } m \to \infty .$

For small enough $\varepsilon > 0$ and for $\{v_m\}_{m \in \mathbb{N}}$ given in (22) we then let

$$w_m^\varepsilon := \begin{cases} v_m & \text{on } B_R(x_0) \\ \hat{u}_m + \varepsilon^{-1}(u_m - \hat{u}_m)(|x| - R) & \text{on } T_\varepsilon - B_R(x_0) \\ u_m & \text{on the rest of } \Omega \end{cases} .$$

Again w_m^ε is seen to be of class $W_1^1(\Omega; \mathbb{R}^N)$ and we have

$$
\begin{aligned}
J[w_m^\varepsilon; \Omega] &= J[v_m; B_R(x_0)] + \int_{T_\varepsilon - B_R(x_0)} f(\nabla w_m^\varepsilon) \, dx \\
&\quad + \int_{\Omega \cap [|x - x_0| > R + \varepsilon]} f(\nabla u_m) \, dx \\
&= J[v_m; B_R(x_0)] + \int_{\Omega - B_R(x_0)} f(\nabla u_m) \, dx \\
&\quad + \int_{T_\varepsilon - B_R(x_0)} f(\nabla w_m^\varepsilon) \, dx - \int_{T_\varepsilon - B_R(x_0)} f(\nabla u_m) \, dx .
\end{aligned}
$$
(25)

As before we deduce from (20) the existence of sequences $\varepsilon_k \to 0$ and $m_k \to \infty$ such that

$$
\int_{T_{\varepsilon_k} - B_R(x_0)} |\nabla u_m| \, dx \le \frac{2}{k} \quad \text{and} \quad \int_{T_{\varepsilon_k} - B_R(x_0)} |\nabla \hat{u}_m| \, dx \le \frac{2}{k}
$$

for all $m \ge m_k$, for all $k \in \mathbb{N}$ and we get

$$
\int_{T_{\varepsilon_k} - B_R(x_0)} |\nabla w_m^{\varepsilon_k}| \, dx \le \frac{6}{k} + \frac{1}{\varepsilon_k} \int_{T_{\varepsilon_k} - B_R(x_0)} |u_m - \hat{u}_m| \, dx \le \frac{6}{k} + \varepsilon_k ,
$$

again for all $k \in \mathbb{N}$ and for all $m \ge m_k$. Thus decomposition (25) gives together with (22) for m_k chosen sufficiently large

$$
\begin{aligned}
J[w_m^{\varepsilon_k}; \Omega] &\le J[v_m; B_R(x_0)] + \int_{\Omega - B_R(x_0)} f(\nabla u_m) \, dx + \beta_k \\
&\le \liminf_{m \to \infty} J[u_m; B_R(x_0)] + \int_{\Omega - B_R(x_0)} f(\nabla u_m) \, dx + 2\beta_k
\end{aligned}
$$

for any $k \in \mathbb{N}$, for all $m \ge m_k$ and with a sequence $\beta_k \to 0$ as $k \to \infty$. Again enlarging m_k if necessary we may assume that

$$
\begin{aligned}
J[w_{m_k}^{\varepsilon_k}; \Omega] &\le \liminf_{m \to \infty} J[u_m; B_R(x_0)] + \liminf_{m \to \infty} J[u_m; \Omega - B_R(x_0)] + 3\beta_k \\
&\le \liminf_{m \to \infty} J[u_m; \Omega] + 3\beta_k .
\end{aligned}
$$

Setting $w_k = w_{m_k}^{\varepsilon_k}$ the lemma is proved observing that as in (24) the definition of w_k also implies L^1 convergence on the whole domain Ω. Moreover, it is clear that we may assume v). $\quad \square$

C

Brief comments on steady states of generalized Newtonian fluids

As one application of the above discussed methods let us give some short comments on steady states of generalized Newtonian fluids. Here, a non-uniform ellipticity condition fits into the main line of this monograph. On the other hand, solenoidal (divergence-free) vector-fields are considered and ∇u has to be replaced by its symmetric part $\varepsilon(u)$.

Besides of several technical difficulties (which we do not want to discuss) this, in particular, gives rise to a challenging open problem: do we expect that a structure condition $f(\varepsilon) = g(|\varepsilon|^2)$ in general is sufficient to imply full regularity of solutions in the above mentioned setting?

Without going into the proofs let us give a short introduction to the results of [BF10] and [ABF] – the main ideas are closely related to the arguments of Section 3.

The stationary flow of an incompressible generalized Newtonian fluid is considered in a bounded Lipschitz domain $\Omega \subset \mathbb{R}^n$, $n = 2$ or $n = 3$. To be precise, we are looking for a velocity field $u \colon \Omega \to \mathbb{R}^n$ solving the following system of nonlinear partial differential equations

$$\left.\begin{array}{c} -\mathrm{div}\left\{ T(\varepsilon(u)) \right\} + u^k \dfrac{\partial u}{\partial x_k} + \nabla \pi = g \quad \text{in } \Omega \,, \\[2mm] \mathrm{div}\, u = 0 \quad \text{in } \Omega \,, \qquad u = 0 \quad \text{on } \partial\Omega \,. \end{array}\right\} \tag{1}$$

Here, π is the a priori unknown pressure function and $g \colon \Omega \to \mathbb{R}^n$ represents a system of volume forces. The tensor T is assumed to be the gradient of some (convex) potential $f \colon \mathbb{S} \to [0, \infty)$ which is of class C^2 on the space $\mathbb{S}$ of all symmetric matrices. Again we adopt the convention of summation over repeated indices running from 1 to n, moreover, for functions $v \colon \Omega \to \mathbb{R}^n$ we let

$$\varepsilon(v)(x) = \frac{1}{2}\left(\partial_i v^j + \partial_j v^i\right)(x) \in \mathbb{S} \,.$$

In the case $f(\varepsilon) = |\varepsilon|^2$, the system (1) reduces to the Dirichlet (no-slip) boundary value problem for the stationary Navier-Stokes system, for an overview on existence and regularity results we refer to the classical monograph [La] of Ladyzhenskaya or, more recently, to the monographs [Ga] of Galdi, where also the history of the problem is outlined in great detail.

So-called power-law models are investigated for example in [KMS]: for some exponent $1 < p < \infty$, f is assumed to satisfy

$$\lambda(1 + |\varepsilon|^2)^{\frac{p-2}{2}}|\sigma|^2 \leq D^2 f(\varepsilon)(\sigma, \sigma) \leq \Lambda(1 + |\varepsilon|^2)^{\frac{p-2}{2}}|\sigma|^2 \tag{2}$$

for all ε, $\sigma \in \mathbb{S}$ and with positive constants λ, Λ. As above, (2) implies that f is of p-growth, moreover, the first inequality in (2) implies strict convexity of f. Then, if $f(\varepsilon) = F(|\varepsilon|^2)$ (which is reasonable from the physical point of view) Kaplický, Málek and Stará discuss the two-dimensional case with the following results: if $p > 3/2$, then the problem (1) admits a solution which is of class $C^{1,\alpha}$ up to the boundary, whereas for $p > 6/5$ the problem (1) has a solution being $C^{1,\alpha}$-regular in the interior of Ω. Here of course the volume force term g is sufficiently smooth.

Suppose for the moment that the flow is also slow. Then in (1) the convective term $(\nabla u)u = u^k \partial_k u$ can be neglected, and (1) reduces to a generalized version of the classical Stokes problem. In the monograph [FuS2], a variational approach towards (1) for various classes of dissipative potentials f is described leading to existence and also (partial) regularity results in the absence of the convective term. Very recently these investigations were extended in [BF10] to the case of non-uniformly elliptic potentials which means that (2) is replaced by the condition

$$\lambda(1 + |\varepsilon|^2)^{\frac{p-2}{2}}|\sigma|^2 \leq D^2 f(\varepsilon)(\sigma, \sigma) \leq \Lambda(1 + |\varepsilon|^2)^{\frac{q-2}{2}}|\sigma|^2 \tag{3}$$

with exponents $1 < p \leq q < \infty$, $q \geq 2$ and for all ε, $\sigma \in \mathbb{S}$. From (3) it easily follows that f is of upper growth rate q, a lower bound for $f(\varepsilon)$ can be given in terms of $|\varepsilon|^p$ (recall Remark 3.5).

Examples of potentials f satisfying (3) are given in [BF10] following the lines of Section 3. Moreover, it is shown in this paper that weak local solutions of (1) (with $(\nabla u)u = 0$!) under condition (3) are $C^{1,\alpha}$ in the interior of Ω if $n = 2$, $q = 2$, and partially $C^{1,\alpha}$ if $n = 3$, provided that we impose the bound $q < p(1 + 2/n)$.

The objective of [ABF] is to study the anisotropic (with respect to the ellipticity condition) situation (3) for a non-vanishing convective term $(\nabla u)u$. To be more precise let us assume that

$$g \in L^\infty(\Omega; \mathbb{R}^n) . \tag{4}$$

Remark C.1. *For the sake of technical simplicity we just assume that the volume forces are bounded functions. Of course our results are also valid under the weaker assumption $g \in L^{t(p)}(\Omega; \mathbb{R}^n)$, whenever $t(p)$ is chosen sufficiently large.*

In order to get a weak form of (1) we multiply the first line of (1) with $\varphi \in C_0^\infty(\Omega; \mathbb{R}^n)$, $\operatorname{div}\varphi = 0$, and obtain after an integration by parts (using $\operatorname{div}\varphi = 0$)

$$\int_\Omega Df(\varepsilon(u)) : \varepsilon(\varphi) \, dx - \int_\Omega u \otimes u : \varepsilon(\varphi) \, dx = \int_\Omega g \cdot \varphi \, dx \, , \qquad (5)$$

where $u \otimes v := (u^i v^k)$. Thus, we have to solve the equation (5) together with $\operatorname{div} u = 0$ in Ω, $u = 0$ on $\partial\Omega$. A priori it is not clear to which space a weak solution should belong, therefore we give an existence proof by using an approximation procedure. To this purpose we again let $(0 < \delta < 1)$

$$f_\delta(\varepsilon) := \delta(1 + |\varepsilon|^2)^{\frac{q}{2}} + f(\varepsilon) \, , \qquad \varepsilon \in S \, ,$$

and consider the problem

$$\left. \begin{array}{l} \text{to find } u_\delta \in \overset{\circ}{W}{}^1_q(\Omega; \mathbb{R}^n), \, \operatorname{div} u_\delta = 0, \text{ such that} \\[2mm] \displaystyle\int_\Omega DF_\delta(\varepsilon(u_\delta)) : \varepsilon(\varphi) \, dx - \int_\Omega u_\delta \otimes u_\delta : \varepsilon(\varphi) \, dx = \int_\Omega g \cdot \varphi \, dx \\[2mm] \text{for all } \varphi \in C_0^\infty(\Omega; \mathbb{R}^n), \, \operatorname{div}\varphi = 0. \end{array} \right\} \qquad (5_\delta)$$

Then we have

Theorem C.2. *Let f satisfy (3) with $1 < p < 2 \le q < \infty$, and consider the data given in (4). Then (5_δ) has at least one weak solution $u_\delta \in \overset{\circ}{W}{}^1_q(\Omega; \mathbb{R}^n)$. Moreover, we have*

$$\sup_{0 < \delta < 1} \|u_\delta\|_{W^1_p(\Omega; \mathbb{R}^n)} < \infty \, .$$

This result is well known and follows from a familiar fixed point argument (see [La]). We remark that by construction u_δ turns out to be the minimizer of the energy

$$J_\delta[w] := \int_\Omega f_\delta(\varepsilon(w)) \, dx - \int_\Omega u_\delta \otimes u_\delta : \varepsilon(w) \, dx - \int_\Omega g \cdot w \, dx \qquad (6)$$

within the class

$$\overset{\circ}{W}{}^1_q(\Omega; \mathbb{R}^n) \cap \operatorname{Kern}(\operatorname{div}) \, .$$

The first observation of [ABF] concerns uniform higher integrability properties of the sequence $\{u_\delta\}$.

Theorem C.3. *Suppose that the assumptions of Theorem C.2 are satisfied and that we have in addition*

$$p > \begin{cases} \frac{6}{5} & \text{in the case } n = 2 \, , \\[2mm] \frac{9}{5} & \text{in the case } n = 3 \, , \end{cases}$$

$$q < p\frac{n+2}{n} \, .$$

Then the functions u_δ are of class $W^1_{\tilde{q},loc}(\Omega; \mathbb{R}^n)$ uniformly with respect to δ, where $\tilde{q} = 3p$ in the case $n = 3$, and where we may choose any finite number $\tilde{q}$ in the case $n = 2$.

As in the sections 2–6, one main ingredient of the proof is a Caccioppoli-type inequality being valid for the approximative solutions u_δ.

The information of Theorem C.3 is used in the next step to pass to the limit, more precisely we have

Theorem C.4. *Let the assumptions of Theorem C.3 hold and fix any weak W^1_p-cluster point of the sequence $\{u_\delta\}$, i.e.*

$$u_\delta \overset{\delta \to 0}{\rightharpoonup} \bar{u} \quad in \ W^1_p(\Omega; \mathbb{R}^n)$$

for some sequence $\delta = \delta_k$ going to zero. Then $\bar{u}$ is of class $W^1_{\tilde{q},loc}(\Omega; \mathbb{R}^n)$, where $\tilde{q}$ is given in Theorem C.3. Moreover, $Df(\varepsilon(\bar{u}))$ locally is of class $W^1_{q/(q-1)}$ and $\bar{u}$ is a solution of the problem (5) being Hölder continuous in the interior of Ω on account of Theorem C.3.

Remark C.5.

i) *In [FMS] the method of Lipschitz approximations of Sobolev functions leads to the existence of solutions for problem (5) in the case of power-law models provided that $p > 6/5$ (see also [FrM]). Note that the convective term in (5) is well defined (in dimension $n = 3$) since the condition $p > 6/5$ implies the solution to be of class L^2. It is not evident, how to apply this method to the non-standard models under consideration (see, for instance, formula (42) of [FrM]). Theorem C.4 just relies on the a priori estimates of Theorem C.3.*

ii) *Note that, in contrast to Section 3, a global regularization is done. This is discussed in detail in [BF11].*

As to the regularity properties of the particular solution $\bar{u}$ from above, the considerations of [ABF] are restricted to the case $q = 2$.

Theorem C.6. *Under the assumptions and with the notation of Theorem C.4 we consider the case $n = 3$, $q = 2$. Then $\bar{u}$ is partially of class $C^{1,\alpha}$, i.e. there is an open set Ω_0 of full Lebesgue measure, $|\Omega - \Omega_0| = 0$, such that $\bar{u} \in C^{1,\alpha}(\Omega_0; \mathbb{R}^3)$.*

Remark C.7. *As in [BF10] it is possible to give a variant of Theorem C.6 for the case $q > 2$ together with $q < 5p/3$.*

Theorem C.8. *Under the assumptions and with the notation of Theorem C.4 we consider the case $n = 2$, $q = 2$. Then $\bar{u}$ has locally Hölder continuous first derivatives, i.e. $\bar{u} \in C^{1,\alpha}(\Omega; \mathbb{R}^2)$.*

Remark C.9. *It would also be desirable to give global variants of the above results, for example to prove higher integrability of $\nabla \bar{u}$ up to the boundary. Then, under suitable smallness conditions, some results on unique solvability extend to our non-uniformly elliptic problems. The idea to establish a theorem of this kind is standard and, for instance, presented in [La], p. 118. The main difficulty in the case of non-uniform ellipticity is to handle potentials with lower growth rate $p < 2$.*

D

Notation and conventions

The set $\Omega \subset \mathbb{R}^n$, $n \geq 2$, always denotes a bounded Lipschitz domain. For $N \geq 1$, the reader is assumed to be familiar with the classical

- Hölder spaces $C^{k,\alpha}(\Omega; \mathbb{R}^N)$,
- Lebesgue spaces $L^p(\Omega; \mathbb{R}^N)$

as well as with the notion of the

- Sobolev spaces $W_p^m(\Omega; \mathbb{R}^N)$, $\overset{\circ}{W}_p^m(\Omega; \mathbb{R}^N)$.

Here we follow the definitions and notation as introduced in [GT], in particular local variants are denoted by

- $L_{loc}^p(\Omega; \mathbb{R}^N)$, etc.

Moreover, the target space is not indicated whenever $N = 1$. The basics on

- Orlicz-Sobolev spaces $W_F^1(\Omega; \mathbb{R}^N)$, $\overset{\circ}{W}_F^1(\Omega; \mathbb{R}^N)$

are shortly outlined in Section 3.1, for more details we refer to [Ad]. Now let $w \in L^1(\Omega; \mathbb{R}^N)$. Then w is called a

- function of bounded variation in Ω, $w \in BV(\Omega; \mathbb{R}^N)$,

if the distributional derivative is representable by a finite Radon measure in Ω, i.e. for some $\mathbb{R}^{nN}$-valued measure $(\partial_\alpha w^i)_{1 \leq \alpha \leq n}^{1 \leq i \leq N}$ we have

$$\int_\Omega w \operatorname{div} \varphi \, dx = \int_\Omega w^i \operatorname{div} \varphi^i \, dx = - \int_\Omega \varphi_\alpha^i \nabla_\alpha w^i$$

$$= - \int_\Omega \varphi : \nabla w \quad \text{for any } \varphi \in C_0^1(\Omega; \mathbb{R}^{nN}).$$

Here and in the following we have the conventions: $\operatorname{div} \varphi = \left(\sum_{\alpha=1}^n \partial_\alpha \varphi^i \right) \in \mathbb{R}^N$. Summation is always assumed with respect to repeated indices – for Latin indices the sum is taken over $i = 1, \ldots, N$, for Greek indices this is done with repect to $\alpha = 1, \ldots, n$ – and the scalar product in $\mathbb{R}^N$ is not highlighted, whereas we take the symbol "$:$" for the standard scalar product in $\mathbb{R}^{nN}$. Moreover, with a slight abuse of notation, derivatives usually are denoted by "∇" – the precise meaning will always be evident by the context.

The total variation of w is given by

$$\int_\Omega |\nabla w| = \sup \left\{ \int_\Omega w \operatorname{div} \varphi \, dx : \varphi \in C_0^1(\Omega; \mathbb{R}^{nN}), \ |\varphi(x)| \leq 1 \ \text{on} \ \Omega \right\},$$

and a L^1-function is seen to be of class $BV(\Omega; \mathbb{R}^N)$ if and only if the total variation is finite. For these definitions and a variety of details we refer to [Giu2] and [AFP].

To consider functionals defined on measures (see [AFP], Section 2.6), let f: $\mathbb{R}^{nN} \to [0, \infty)$ be strictly convex, continuous, of linear growth with $f(0) = 0$ (of course, this assumption is too strong but always satisfied in our studies). The recession function $f_\infty \colon \mathbb{R}^{nN} \to \mathbb{R}$ is given by

$$f_\infty(Z) := \limsup_{t \uparrow \infty} \frac{f(tZ)}{t} = \lim_{t \uparrow \infty} \frac{f(tZ)}{t} \quad \text{for any} \ Z \in \mathbb{R}^{nN} .$$

Obviously, f_∞ is positively homogeneous of degree one.

Now, any $\mathbb{R}^{nN}$-valued measure μ in Ω will always be decomposed with respect to the Lebesgue measure. Then μ^a denotes the absolutely continuous part, whereas μ^s denotes the singular part and $\mu^s / |\mu^s|$ is the symbol for the Radon-Nikodym derivative. With this notation we may define (see [AFP], formula (2.26))

$$G(\mu) := \int_\Omega f\big(\mu^a(x)\big) \, dx + \int_\Omega f_\infty \left(\frac{\mu^s}{|\mu^s|}(x) \right) d|\mu^s|(x) . \tag{1}$$

Finally, if f^* denotes the conjugate of f,

$$f^*(Q) := \sup_{Z \in \mathbb{R}^{nN}} \{ Z : Q - f(Z) \} \quad \text{for any} \ Q \in \mathbb{R}^{nN} ,$$

then we may let for any Borel set $B \subset \Omega$

$$F_B(\mu) := \sup_{\varkappa \in C_0^\infty(\Omega; \mathbb{R}^{nN}), \, f^* \circ \varkappa \in L^1(\Omega)} \left\{ \int_\Omega 1_B \varkappa \mu - \int_\Omega f^*(\varkappa) \, dx \right\} . \tag{2}$$

Note that by Proposition 1.2, [DT], both viewpoints (1) and (2) coincide.

It remains to mention three conventions which are not restated each time:

- if we do not give further comments, then any ball under consideration is assumed to be open and compactly contained in Ω;
- if necessary (and possible), we usually pass to subsequences without relabeling;
- positive constants are also not relabeled and usually just denoted by c, not necessarily being the same in any two occurrencies. Moreover, only relevant dependences are highlighted.

References

[Ad] Adams, R. A., Sobolev spaces. Academic Press, New York-San Francisco-London 1975.

[Alm] Almgren, F. J.,Jr., Existence and regularity almost everywhere of solutions to elliptic variational problems among surfaces of varying topological type and singularity structure. Annals of Math. 87 (1968), 321–391.

[Alt] Alt, H.W., Lineare Funktionalanalysis. Springer, Berlin-Heidelberg 1985.

[ABF] Apouchkinskaya, D., Bildhauer, M., Fuchs, M., Steady states of anisotropic generalized Newtonian fluids. Submitted.

[AD] Ambrosio, L., Dal Maso, G., On the relaxation in $BV(\Omega_{\mid} \mathbb{R}^m)$ of quasiconvex integrals. J. Funct. Anal. 109 (1992), 76–97.

[AF1] Acerbi, E., Fusco, N., A regularity theorem for minimizers of quasiconvex integrals. Arch. Rat. Mech. Anal. 99 (1987), 261–281.

[AF2] Acerbi, E., Fusco, N., Local regularity for minimizers of non convex integrals. Ann. Sc. Norm. Super. Pisa, Cl. Sci., IV. Ser. 16, no. 4 (1989), 603–636.

[AF3] Acerbi, E., Fusco, N., Regularity for minimizers of non-quadratic functionals: the case $1 < p < 2$. J. Math. Anal. Appl. 140 (1989), 115–135.

[AF4] Acerbi, E., Fusco, N., Partial regularity under anisotropic (p, q) growth conditions. J. Diff. Equ. 107, no. 1 (1994), 46–67.

[AFP] Ambrosio, L., Fusco, N., Pallara, D., Functions of bounded variation and free discontinuity problems. Oxford Science Publications, Clarendon Press, Oxford 2000.

[AG1] Anzellotti, G., Giaquinta, M., Funzioni BV e tracce. Rend. Sem. Mat. Padova 60 (1978), 1–21.

[AG2] Anzellotti, G., Giaquinta, M., Convex functionals and partial regularity. Arch. Rat. Mech. Anal. 102 (1988), 243–272.

[Bi1] Bildhauer, M., A uniqueness theorem for the dual problem associated to a variational problem with linear growth. Zap. Nauchn. Sem. St.-Petersburg Odtel. Math. Inst. Steklov (POMI) 271 (2000), 83–91.

[Bi2] Bildhauer, M., A note on degenerate variational problems with linear growth. Z. Anal. Anw. 20, no. 3 (2001), 589–598.

[Bi3] Bildhauer, M., Convex variational problems with linear, nearly linear and/or anisotropic growth conditions. Habilitationsschrift, Saarland University, 2001.

208   References

[Bi4]   Bildhauer, M., A priori gradient estimates for bounded generalized so-
          lutions of a class of variational problems with linear growth. J. Convex
          Anal. 9 (2002), 117–137.
[Bi5]   Bildhauer, M., Two dimensional variational problems with linear growth.
          Manus. Math. 110 (2003), 325–342.
[Bi6]   Bildhauer, M., Convex variational integrals with mixed anisotropic lin-
          ear/superlinear growth conditions. Submitted.
[Bu]   Buttazzo, G., Semicontinuity, relaxation and integral representation in
          the calculus of variations. Pitman Res. Notes in Math., Longman, Harlow
          1989.
[BF1]   Bildhauer, M., Fuchs, M., Regularity for dual solutions and for weak clus-
          ter points of minimizing sequences of variational problems with linear
          growth. Zap. Nauchn. Sem. St.-Petersburg Odtel. Math. Inst. Steklov
          (POMI) 259 (1999), 46–66.
[BF2]   Bildhauer, M., Fuchs, M., Partial regularity for variational integrals with
          (s, μ, q)-growth. Calc. Var. 13 (2001), 537–560.
[BF3]   Bildhauer, M., Fuchs, M., On a class of variational integrals with linear
          growth satisfying the condition of μ-ellipticity. To appear in Rend. Mat. e
          delle sue appl.
[BF4]   Bildhauer, M., Fuchs, M., Relaxation of convex variational problems with
          linear growth defined on classes of vector-valued functions. Algebra and
          Analiz 14, no. 1, (2002), 26–45.
[BF5]   Bildhauer, M., Fuchs, M., Partial regularity for a class of anisotropic vari-
          ational integrals with convex hull property. Asymp. Anal. 32 (2002), 293–
          315.
[BF6]   Bildhauer, M., Fuchs, M., Twodimensional anisotropic variational prob-
          lems. Calc. Var. 16 (2003), 177–186.
[BF7]   Bildhauer, M., Fuchs, M., Elliptic variational problems with nonstandard
          growth. In: International Mathematical Series, Vol. 1, Nonlinear problems
          in mathematical physics and related topics I, in honor of Prof. O.A. La-
          dyzhenskaya. By T. Rozhkovskaya, Novosibirsk, Russia, 2002 (in Russian),
          49–62. By Kluwer/Plenum Publishers, 2002 (in English), 53–66.
[BF8]   Bildhauer, M., Fuchs, M., Convex variational problems with linear growth.
          In: Geometric analysis and nonlinear partial differential equations. By
          S. Hildebrandt and H. Karcher, Springer, Berlin-Heidelberg-New York
          2003, 327–344.
[BF9]   Bildhauer, M., Fuchs, M., Interior regularity for free and constrained local
          minimzers of variational integrals under general growth and ellipticity
          conditions. Zap. Nauchn. Sem. St.-Petersburg Odtel. Math. Inst. Steklov
          (POMI) 288 (2002), 79–99.
[BF10]   Bildhauer, M., Fuchs, M., Variants of the Stokes problem: the case of
          anisotropic potentials. J. Math. Fluid Mech. 5 (2003), 1–39.
[BF11]   Regularization of some variational problems with applications to general-
          ized Newtonian fluids. Submitted.
[BFM]   Bildhauer, M., Fuchs, M., Mingione, G., A priori gradient bounds and
          local $C^{1,\alpha}$-estimates for (double) obstacle problems under nonstandard
          growth conditions. Z. Anal. Anw. 20, no.4 (2001), 959–985.
[BGM]   Bombieri, E., DeGiorgi, E., Miranda, M., Una maggiorazione
          a priori relativa alle ipersuperfici minimali non parametriche.
          Arch. Rat. Mech. Anal. 32 (1969), 255–267.

[GiuM1] Giusti, E., Miranda, M., Sulla regolarità delle soluzioni deboli di una classe di sistemi ellittici quasi-lineari. Arch. Rat. Mech. Anal. 31 (1968), 173–184.

[GiuM2] Giusti, E., Miranda, M., Un esempio di soluzioni discontinue per un problema di minimo relativo ad un integrale regolare del calcolo delle variazioni. Boll. UMI 2 (1968), 1–8.

[GMS1] Giaquinta, M., Modica, G., Souček, J., Functionals with linear growth in the calculus of variations. Comm. Math. Univ. Carolinae 20 (1979), 143–172.

[GMS2] Giaquinta, M., Modica, G., Souček, J., Cartesian currents in the calculus of variations I & II. Ergebnisse der Mathematik und ihrer Grenzgebiete, vol. 37 & 38, Springer, Berlin-Heidelberg 1998.

[GS] Goffman, C., Serrin, J., Sublinear functions of measures and variational integrals. Duke Math. J. 31 (1964), 159–178.

[GT] Gilbarg, D., Trudinger, N.S., Elliptic partial differential equations of second order. Grundlehren der math. Wiss. 224, second ed., revised third print., Springer, Berlin-Heidelberg-New York 1998.

[KMS] Kaplický, P., Málek, J., Stará, J., $C^{1,\alpha}$-solutions to a class of nonlinear fluids in two dimensions – stationary Dirichlet problem. Zap. Nauchn. Sem. St.-Petersburg Odtel. Math. Inst. Steklov (POMI) 259 (1999), 89–121.

[KS] Kinderlehrer, D., Stampacchia, G., An introduction to variational inequalities and their applications. Academic Press, New York-San Francisco-London 1980.

[La] Ladyzhenskaya, O.A., The mathematical theory of viscous incompressible flow. Gordon and Breach, 1969.

[Lie1] Lieberman, G., Regularity of solutions to some degenerate double obstacle problems. Indiana Univ. Math. J. 40, No. 3 (1991), 1009–1028.

[Lie2] Lieberman, G., Gradient estimates for a class of elliptic systems. Ann. Mat. Pura Appl., IV. Ser. 164 (1993), 103–120.

[Lin] Lindqvist, P., Regularity for the gradient of the solution to a nonlinear obstacle problem with degenerate ellipticity. Nonlinear Anal. 12 (1988), 1245–1255.

[LU1] Ladyzhenskaya, O.A., Ural'tseva, N.N., Linear and quasilinear elliptic equations. Nauka, Moskow, 1964. English translation: Academic Press, New York 1968.

[LU2] Ladyzhenskaya, O.A., Ural'tseva, N.N., Local estimates for gradients of solutions of non-uniformly elliptic and parabolic equations. Comm. Pure Appl. Math. 23 (1970), 667–703.

[LU3] Ladyzhenskaya, O.A., Ural'tseva, N.N., Local estimates for the gradients of solutions to the simplest regularization of a class of nonuniformly elliptic equations. Zap. Nauchn. Sem. St.-Petersburg Odtel. Math. Inst. Steklov (POMI) 213 (1994), 75–92 (in Russian). English translation: J. Math. Sci. 84 (1997), 862–872.

[Ma1] Marcellini, P., Approximation of quasiconvex functions, and lower semicontinuity of multiple integrals. Manus. Math. 51 (1985), 1–28.

[Ma2] Marcellini, P., Regularity of minimizers of integrals of the calculus of variations with non standard growth conditions. Arch. Rat. Mech. Anal. 105 (1989), 267–284.

212   References

[Ma3]   Marcellini, P., Regularity and existence of solutions of elliptic equations with (p, q)-growth conditions. J. Diff. Equ. 90 (1991), 1–30.

[Ma4]   Marcellini, P., Regularity for elliptic equations with general growth conditions. J. Diff. Equ. 105 (1993), 296–333.

[Ma5]   Marcellini, P., Everywhere regularity for a class of elliptic systems without growth conditions. Ann. Scuola Norm. Sup. Pisa 23 (1996), 1–25.

[Ma6]   Marcellini, P., Regularity for some scalar variational problems under general growth conditions. J. Optim. Theory Appl. 90 (1996), 161–181.

[Ma7]   Marcellini, P., General growth conditions and regularity. In: Variational methods for discontinuous structures (Como 1994), 111–118, Birkhäuser, Basel 1996.

[Mi]   Miranda, M., Disequaglianze di Sobolev sulle ipersuperfici minimali. Rend. Sem. Mat. Univ. Padova 38 (1967), 69–79.

[Mor1]   Morrey, C. B., Multiple integrals in the calculus of variations. Grundlehren der math. Wiss. in Einzeldarstellungen 130, Springer, Berlin-Heidelberg-New York 1966.

[Mor2]   Morrey, C. B., Partial regularity results for nonlinear elliptic systems. J. Math. Mech. 17 (1968), 649–670.

[Mos]   Moser, J., A new proof of DeGiorgi's theorem concerning the regularity problem for elliptic differential equations. Comm. Pure Appl. Math. 13 (1960), 457–468.

[MM]   Massari, U., Miranda, M., Minimal surfaces of codimension one. Noth-Holland Mathematics Studies 91, North-Holland, Amsterdam 1984.

[MS]   Mingione, G., Siepe, F., Full $C^{1,\alpha}$-regularity for minimizers of integral functionals with $L \log L$ growth. Z. Anal. Anw. 18 (1999), 1083–1100.

[MiZ]   Michael, J., Ziemer W., Interior regularity for solutions to obstacle problems. Nonlinear Anal. 10 (1986), 1427–1448.

[MuZ]   Mu, J., Ziemer, W. P., Smooth regularity of solutions of double obstacle problems involving degenerate elliptic equations. Comm. P.D.E. 16, nos. 4–5 (1991), 821–843.

[Na]   Nash, J., Continuity of solutions of parabolic and elliptic equations. Amer. J. Math. 80 (1958), 931–954.

[Ne]   Nečas, J., Example of an irregular solution to a nonlinear elliptic system with analytic coefficients and conditions of regularity. In: Theory of Non Linear Operators, Abhandlungen Akad. der Wissen. der DDR, Proc. of a summer school held in Berlin (1975), 197–206, Berlin 1977.

[PS]   Passarelli Di Napoli, A., Siepe, F., A regularity result for a class of anisotropic systems. Rend. Ist. Mat. Univ. Trieste 28, No.1-2 (1996), 13–31.

[Re]   Reshetnyak, Y., Weak convergence of completely additive vector functions on a set. Sibirsk. Maz. Ž., 9 (1968), 1386–1394 (in Russian). English translation: Sib. Math. J 9 (1968), 1039–1045.

[Ro]   Rockafellar, T., Convex analysis. Princeton University Press, Princeton 1970.

[Sch]   Schwartz, J.T., Nonlinear functional analysis. Gordon and Breach Science Publishers, New York-London-Paris, 1969.

[Se1]   Seregin, G., Variational-difference scheme for problems in the mechanics of ideally elastoplastic media. Zh. Vychisl. Mat. Fiz. 25 (1985), 237–352 (in Russian). English translation: U.S.S.R Comp. Math. and Math. Phys. 25 (1985), 153–165.

[Se2]   Seregin, G., Differentiability of local extremals of variational problems in the mechanics of perfect elastoplastic media. Differentsial'nye Uravneniya 23 (11) (1987), 1981–1991 (in Russian). English translation: Differential Equations 23 (1987), 1349–1358.

[Se3]   Seregin, G.A., On differential properties of extremals of variational problems arising in plasticity theory. Differentsial'nye Uravneniya 26 (1990), 1033–1043 (in Russian). English translation: Differential Equations 26 (1990), 756–766

[Se4]   Seregin, G.A., Differential properties of solutions of variational problems for functionals with linear growth. Problemy Matematicheskogo Analiza, Vypusk 11, Isazadel'stvo LGU (1990), 51–79 (in Russian). English translation: J. Soviet Math. 64 (1993), 1256–1277.

[Se5]   Seregin, G., Differentiability properties of weak solutions of certain variational problems in the theory of perfect elastoplastic plates. Appl. Math. Optim. 28 (1993), 307–335.

[Se6]   Seregin, G., Twodimensional variational problems in plasticity theory. Izv. Russian Academy of Sciences 60 (1996), 175–210 (in Russian). English translation: Izvestiya Mathematics 60.1 (1996), 179–216.

[St]   Stampacchia, G., Le problème de Dirichlet pour les équations elliptiques du second ordre á coefficients discontinus. Ann. Inst. Fourier Grenoble 15.1 (1965), 189–258.

[ST]   Strang, G., Temam, R., Duality and relaxations in the theory of plasticity. J. Méchanique 19 (1980), 1–35.

[SY]   Šverák, V., Yan, X., A singular minimizer of a smooth strongly convex functional in three dimensions. Calc. Var. 10 (2000), 213–221.

[Ta]   Talenti, G., Boundedness of minimizers. Hokkaido Math. J. 19 (1990), 259–279.

[Te]   Temam, R., Approximation de fonctions convexes sur un espace de mesures et applications. Cana. Math. Bull. 25 (1982), 392–413.

[Uh]   Uhlenbeck, K., Regularity for a class of nonlinear elliptic systems. Acta Math. 138 (1977), 219–240.

[Ur]   Ural'tseva. N.N., Quasilinear degenerate elliptic systems. Leningrad Odtel. Mat. Inst. Steklov (LOMI) 7 (1968), 184–222 (in Russian).

[UU]   Ural'tseva, N.N., Urdaletova, A.B., The boundedness of the gradients of generalized solutions of degenerate quasilinear nonuniformly elliptic equations. Vestn. Leningr. Univ. 1983, Mat. Mekh. Astron. no. 4 (1983), 50–56 (in Russian). English translation: Vestn. Leningr. Univ. Math 16 (1984), 263–270.

[Ze]   Zeidler, E., Nonlinear functional analysis and its applications III. Springer, New York-Berlin-Heidelberg-Tokyo 1984.

Index

Printing and Binding: Strauss GmbH, Mörlenbach

4. Lecture Notes are printed by photo-offset from the master-copy delivered in camera-ready form by the authors. Springer-Verlag provides technical instructions for the preparation of manuscripts. Macro-packages in LaTeX2e are available from Springer's web-pages at

http://www.springer.de/math/authors/index.html

Macros in LaTeX2.09 and TeX are available on request at: lnm@springer.de. Careful preparation of the manuscripts will help keep production time short and ensure satisfactory appearance of the finished book. After acceptance of the manuscript authors will be asked to prepare the final LaTeX source files (and also the corresponding dvi- or pdf-file) together with the final printout made from these files. The LaTeX source files are essential for producing the full-text online version of the book

(http://www.springerlink.com/link/service/series/0304/tocs.htm).

The actual production of a Lecture Notes volume takes approximately 12 weeks.

5. Authors receive a total of 50 free copies of their volume, but no royalties. They are entitled to a discount of 33.3 % on the price of Springer books purchased for their personal use, if ordering directly from Springer-Verlag.

6. Commitment to publish is made by letter of intent rather than by signing a formal contract. Springer-Verlag secures the copyright for each volume. Authors are free to reuse material contained in their LNM volumes in later publications: A brief written (or e-mail) request for formal permission is sufficient.

Addresses:

Professor J.-M. Morel, CMLA,
Ecole Normale Supérieure de Cachan,
61 Avenue du Président Wilson, 94235 Cachan Cedex, France
E-mail: Jean-Michel.Morel@cmla.ens-cachan.fr

Professor F. Takens, Mathematisch Instituut,
Rijksuniversiteit Groningen, Postbus 800,
9700 AV Groningen, The Netherlands
E-mail: F.Takens@math.rug.nl

Professor B. Teissier, Université Paris 7
Institut Mathématique de Jussieu, UMR 7586 du CNRS
Equipe "Géométrie et Dynamique", 175 rue du Chevaleret
75013 Paris, France
E-mail: teissier@math.jussieu.fr

Springer-Verlag, Mathematics Editorial, Tiergartenstr. 17,
69121 Heidelberg, Germany,
Tel.: +49 (6221) 487-8410
Fax: +49 (6221) 487-8355
E-mail: lnm@Springer.de